THE AI LITERACY PLAYBOOK FOR HOTELIERS

MICHAEL J. GOLDRICH

Published by Vivander Advisors LLC

ISBN: 979-8-89079-343-0 (hardcover)
ISBN: 979-8-89079-342-3 (paperback)
ISBN: 979-8-89079-332-4 (ebook)

CONTENTS

Part 3: Responsible and Effective Implementation

Part 4: The Human Element in an AI-Augmented Future

Part 5: Advanced AI Systems AND Strategic Vision

Appendices

FOREWORD

AI PERSONAS OF HOSPITALITY PAST

In my book, *Too Many Hats, Too Little Time*, I examined the time demands professionals face and the growing role AI can play in giving time back. The core idea was simple: AI is most powerful when it has a clear job. The more defined the role, the more valuable the support. That is where the concept of AI personas originated. AI personas are virtual team members with titles, like CMO or revenue manager, equipped with domain knowledge, personality, skills, trained prompts, and a specific point of view. These were designed to think in context, act with purpose, and work alongside human teams. The book explored how to build and use them, emphasizing a shift in mindset: moving from automation to collaboration. This represented a change from "How do I use AI for this task?" to "How do I work smarter with my 24/7 subject-matter-expert assistant?"

That perspective got me thinking: *If the great hospitality leaders of the past had access to generative AI, how would they approach today's biggest challenge, the Time Paradox*? This is where being so busy with day-to-day tasks makes it almost impossible to step back and take the time to implement the tools that could relieve you.

This led me to an experiment. What if we could build AI personas of iconic figures in hospitality history modeled after their leadership philosophies, values, and business achievements? So, that is what I did. I created AI personas inspired by pioneers like Conrad Hilton, who saw expansion as a tool for global connection, and J.W. Marriott, who embedded discipline and consistency into every operational detail. Then I posed a simple, modern question to each persona: "If you were leading your hotel company today, how would you use generative AI to address the Time Paradox, being too busy with daily tasks to invest time in tools that could save time?"

Their "responses," generated using prompt frameworks built on their leadership styles, surfaced real, practical insights about how today's hospitality leaders might better approach AI adoption as a strategic partner.

- The AI persona of César Ritz, whose name defines luxury and intuitive service, responded: "Generative AI? An extraordinary tool! I would use it to achieve an even higher level of 'emotional precision' in guest service. Imagine AI helping staff understand individual preferences before guests arrive, crafting perfectly personalized welcome notes, or anticipating unstated needs. This would augment, not replace, the human touch. It would free my team from routine communication, giving them more time for memorable, bespoke moments that define Ritz service. AI could also develop consistent training materials, ensuring every team member understands our exacting standards with less traditional instruction time."
- Next, the AI persona of Ellsworth M. Statler, who brought efficiency and standardized quality to American hotels with his promise of "a bed and a bath for a dollar and a half,"

offered: "AI fits perfectly with my systematic approach. I would use it to streamline administrative toil by automating routine reports, managing inventory with predictive analysis, and creating efficient SOPs easily accessible to all staff. An AI could analyze maintenance logs to predict equipment failures before disrupting guests, saving repair time and service recovery costs. By reducing these operational burdens, my teams would have more time for flawless, efficient guest service, the core of the Statler promise."

- The AI persona of Conrad Hilton, who built a global network balancing brand standard with local character, explained: "An interesting challenge. AI would be invaluable for maintaining that balance. I see it helping adapt global brand messages with greater cultural distinction for diverse markets, a task that takes considerable human time and expertise. AI could also give my general managers tools to rapidly analyze local market trends, property-specific guest feedback, and competitor activities. It allows faster, more informed decisions to effectively customize their hotel's offerings, saving strategic time."
- Then, the AI persona of Kemmons Wilson, the practical founder of Holiday Inn, focused on reliable, comfortable lodging for families, chimed in: "My questions would be simple: How does this AI save my guests or franchisees time and hassle? Can AI make family check-in faster after a long drive? Yes? Good. Can AI help new franchisees train staff on core service standards more quickly than with manuals alone? Yes? Excellent. Can AI help us quickly test different marketing messages to attract families? If it delivers clear, practical benefits by reducing toil for guests and staff and improving efficiency, then I am all for it. It pertains to results."
- Finally, the AI persona of J. Willard Marriott, who built his company on careful attention to detail and operational discipline, stated: "I would integrate AI systematically where

> it enhances precision and empowers our people. Imagine AI creating detailed, easily updatable training modules for every role, ensuring consistency across properties with less repetitive training. AI can also analyze operational data to identify inefficiencies or system-wide improvement opportunities that individual managers might miss. I would focus on AI's potential to improve reliability and consistency, freeing our teams for direct, personal guest engagement, where true hospitality lies."

These AI personas, reflecting their human counterparts' innovative spirit, would likely see today's AI as a powerful tool to extend their established principles of excellence, efficiency, and guest focus. They understood that embracing new tools was vital. This is the mindset this playbook aims to develop: approaching AI with curiosity, strategic focus, and the goal of enhancing human skills and guest satisfaction.

INTRODUCTION

AI ASCENSION

If you are a busy professional in the demanding world of hospitality, you know the feeling: There are never enough hours in the day. You are likely grappling with the relentless demand to do more with less—more personalized guest experiences, more marketing reach, and more operational efficiency, while resources often feel stretched thin. You might be deep in the day-to-day, managing teams, hitting deadlines, and fighting fires. This situation makes it almost impossible to step back and implement the tools that could relieve you.

So, why should you carve out precious time from your packed schedule to explore artificial intelligence through this playbook? Because AI, when understood and applied strategically, is no longer a distant concept. It is a set of practical tools available now that can directly address the persistent pressures you face and help you reclaim your most valuable asset: time.

This playbook is written with your reality in mind. It is designed to cut through the hype and show you, in straightforward terms, how AI can help you in several ways. First, AI can help you reclaim your day and your focus. Imagine AI handling the first draft of those many repetitive and necessary emails or summarizing lengthy reports and meeting transcripts. This frees you and your team for more impactful tasks. The core idea I have explored is simple: AI is useful only when it has a clear job. The more defined the role, the more valuable the support. This approach significantly reduces the toil of the mundane, repetitive tasks that drain energy. The goal is to create more space for joy. This meaningful, creative, and strategic work makes a difference.

Second, AI can elevate service quality and personalization. AI can give your team instant access to insights and information, enabling more consistent and personalized interactions, even when things are hectic. It can be powered by tools to answer common guest inquiries accurately 24/7. It can also help craft perfectly customized communications that anticipate guest needs, augmenting the human touch.

Third, AI can unlock new avenues for growth and efficiency. AI tools can analyze market trends and operational data to help optimize pricing, identify new revenue opportunities, and personalize marketing efforts. Beyond revenue, think of AI predicting potential maintenance issues, optimizing resource schedules, or helping to reduce waste through better forecasting. This contributes to a smoother, more intelligent operation.

Fourth, AI can empower your team. A common apprehension is AI's impact on people. This playbook will consistently show you how AI can be a collaborative partner, an "always ready" assistant designed to take on the tasks that cause frustration and burnout. This frees up talent to apply their uniquely human skills like empathy, critical thinking, and genuine connection more effectively, making their roles more fulfilling.

This journey is about strategically applying intelligent tools to solve the real-world challenges you face daily. This makes your operations smarter, your guest experiences richer, and your team more effective and engaged. This playbook is your guide to understanding how to build AI literacy, cultivate an AI mindset, and lead with confidence into an AI-assisted future in hospitality.

Author's Note:

Like many, my deep examination into generative artificial intelligence began when ChatGPT hit the scene. This technology looked like it could be a panacea, a genuine way to alleviate day-to-day pressures ranging from the hours our front desk teams spend crafting personalized, repetitive welcome emails to the struggle for marketing managers to generate diverse promotional content with limited budgets.

My work at Vivander Advisors is the product of a career spent where hospitality and technology intersect. I began in management consulting and spent two decades on commercial teams across global brands, management companies, and boutique hotels. Over time, my focus has shifted toward something with even greater potential: helping the industry tap into the disruptive power of AI to work smarter, lead better, and grow faster. This pivot led me to a role at a startup dedicated to providing machine learning capabilities and predictive personalization to over 20,000 hotels. Concurrently, I was also shaping future thinking as the program leader for Northwestern University's course on generative AI platforms.

With this background, the arrival of ChatGPT presented an undeniable opportunity. The implications were so compelling that I quickly discussed them with the Hospitality Sales & Marketing Association International (HSMAI) Marketing Association Board in early December of 2022. That dialogue sparked the creation of Vivander Advisors, with its clear mission: to provide AI thought leadership, deliver practical AI education, and help develop effective AI solutions for hoteliers worldwide.

As I spoke with more hoteliers at conferences, in workshops, and master classes, a key insight emerged. When I first started researching how companies were deploying generative AI, the pattern was impossible to ignore. Headlines touted breakthrough stories. Demos dazzled audiences, and tools like ChatGPT and Copilot spread like wildfire across industries. Still, reality told a different story inside many organizations: AI pilots fizzled, platforms sat unused, and expectations quietly deflated. This disconnect reminded me of Nobel laureate Robert Solow's famous 1987 line: "You can see the computer age everywhere except in the productivity statistics." Solow pointed to a mismatch between technological hype and measurable economic output. Back then, despite heavy investment in computers, productivity numbers barely budged. I wrote my master's thesis at Northwestern University on that paradox: Technological progress does not automatically translate into productivity gains. I proposed a solution centered around measurement. Today, generative AI presents a strikingly similar dilemma. AI features in every boardroom deck and startup pitch fail to always appear in the day-to-day output or efficiency of the average workplace. That tension is what I call the Generative AI Productivity Paradox. The gap between AI's significant potential and its underwhelming real-world impact.

The Paradox: Why AI Promises Big but Underwhelms in Practice

On paper, generative AI should change everything. It creates original content instantly, reduces time spent on repetitive tasks, and opens doors to new kinds of work and collaboration. The numbers back this up: McKinsey, BCG, MIT, and Bain all report massive gains in productivity, cost savings, and innovation potential when generative AI is used well. However, most companies have failed to see those gains materialize. According to Gartner, 85% of AI projects fail to deliver business value. Employees are curious but confused. Leaders are optimistic, yet unsure where to begin. So, what is going wrong?

The Causes: Why Generative AI Projects Fall Short

After speaking with leaders, consulting with clients, and reviewing the data, I have found that the root causes are not necessarily technical. They are organizational, behavioral, and strategic.

- **Strategic Misalignment**: Companies prioritize short-term tactics over long-term significant change. Many projects launch without a clear ROI framework or operational objective, making them vulnerable to the first sign of difficulty.
- **Workforce Unreadiness:** Despite nearly universal curiosity about AI, only 5% of companies offer training at scale. Employees lack the skills and confidence to meaningfully use these tools. Resistance, fear of replacement, and poor communication continue to widen this gap.
- **Operational and Technological Gaps**: AI gets bolted onto broken workflows, creating frustration rather than efficiency. Poor data infrastructure and incompatible tools prevent scaling. And automation fails when the process itself is flawed.
- **Unrealistic Expectations:** Leaders imagine overnight revolution and underestimate the learning curve. Without proper change management and ongoing support, early efforts stall and enthusiasm wanes.
- **Lack of AI Literacy:** The absence of understanding how AI works, when to trust it, or how to interact with it effectively leaves most workers in the dark. Ethical issues around bias, transparency, and misuse go unaddressed, reducing confidence in the technology and limiting adoption.

The Solution:
Building AI Literacy + Habit + Strategy

To break the paradox, we need a different playbook, one grounded not in technology alone but in human behavior, learning, and culture. That is why I created the AI Literacy Mindset Program, designed to bridge the gap between AI's potential and the operational realities of hotel environments. This approach's core is a simple and powerful formula: AI Literacy + AI Mindset = Retention & Growth. This entails building AI fluency, confidence, and a shared sense of purpose across every level of the organization.

This is the core idea behind AI ascension. In many companies, AI tools are introduced from the top down. Executives choose a platform and push it out to teams, often with little explanation or training. This approach rarely works. It skips the human part. Real change does not start with tools; it starts with people.

AI ascension flips this script. Instead of forcing advanced technology into workflows that aren't ready, we begin by building AI confidence and curiosity at the front lines. Associates and managers learn what AI is, how to use it, and where it can help them right now, not sometime in the distant future.

Once frontline staff see results and feel capable, something powerful happens. They stop waiting for more direction from above. They start pulling advanced AI into their work. They begin asking for better tools, experimenting with automation, and sharing best practices. This shift from top-down pushing to bottom-up pulling is where the real change happens. That's when AI stops being a project and becomes part of the culture. This is where AI adoption truly begins.

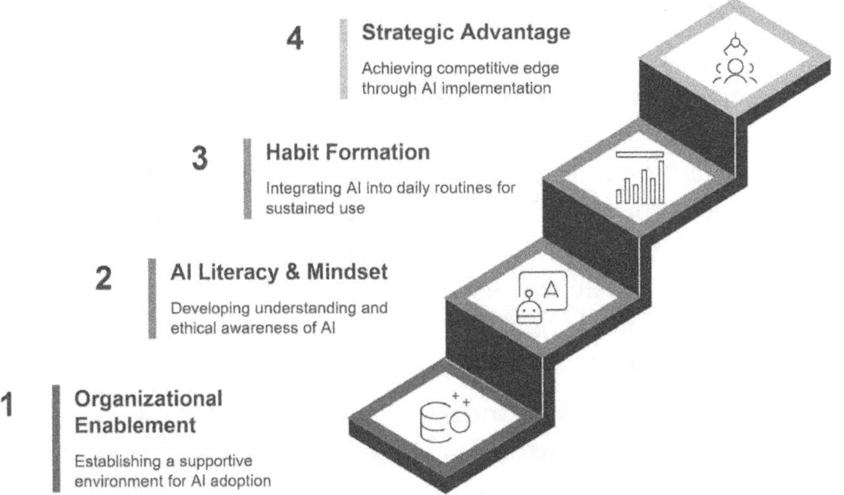

- **Organizational Enablement:** This foundational stage establishes a supportive ecosystem for successful AI adoption. This environment is crucial for the subsequent stages of AI Literacy & Mindset, Habit Formation, and Strategic Advantage to develop and thrive. A key component of this support system is the cultivation of AI champions, enthusiastic volunteers from various departments who help evangelize the benefits and support the adoption of AI in the hotel. They act as peer mentors and first-line support, fostering enthusiasm and practical skill-building from within the team. This supportive structure is built on four key pillars, known as the 4 Ts:
    - Tone from the top
    - Tools to use
    - Training to upskill
    - Time to experiment

For AI adoption to succeed, executives must lead by example, modeling and encouraging AI-driven behaviors. It is essential to provide teams with access to safe and reliable AI platforms that are easy to use. Training should be structured and continuous, rather than a one-time event. Furthermore, teams need to be given the permission and time to explore, experiment, and even fail as they learn and improve. When this foundational support is in place, it creates an environment where frontline associates can build confidence and drive AI adoption from the ground up, ultimately leading the organization toward a strategic advantage

- **AI Literacy & Mindset**: AI adoption starts with understanding. It means knowing what AI is, how it works, and what it can (and cannot) do. It includes ethics and bias awareness. It builds practical skills in prompting and platform use. It connects AI capabilities to real business outcomes and lays the foundation for responsible use, which is critical in a compliance-heavy industry like hospitality.
- **Habit Formation**: Driven by the eleven-by-eleven tipping point. Microsoft's research shows that when people save at least eleven minutes per day with AI and consistently use it for eleven weeks straight, usage shifts from novelty to necessity. That is when AI starts to stick. To build those habits in hospitality, we introduce a daily AI-first mindset. A simple question, asked repeatedly: Can AI help initiate or improve this task? That shift in thinking drives small and consistent gains. In practice, a front office manager might reserve fifteen minutes each week for staff to share AI wins: a task sped up, a guest email drafted, a report started. These repeated micro-moments reinforce new behavior. Throughout this playbook, you will see how sustained, incremental use unlocks real performance gains. The goal is not to train people once and walk away; it is to embed

AI into the operational rhythm of the hotel, one use case, one habit, one step at a time.

- **Strategic Advantage:** When the first three steps—Organizational Enablement, providing the supportive ecosystem, AI Literacy & Mindset furnishing the necessary knowledge and approach, and Habit Formation ensuring consistent application—are strategically integrated into your hotel. Strategic Advantage is characterized by enhanced operational efficiency, elevated guest experiences driven by empowered staff, improved employee engagement through reduced toil, data-driven agility, and a culture of continuous, frontline-driven innovation. This holistic, human-centric approach to AI adoption propels your hotel to the higher levels of AI ascension, establishing a distinct and sustainable market leadership position.

The AI Balancing Act: Marrying Quick Wins with Long-Term Vision

As you integrate artificial intelligence into your hotel's operations, you will encounter a natural dynamic: the allure of quick wins versus the necessity of a long-term, strategic AI vision. Successfully navigating this balance is key to building sustainable AI capabilities and fostering genuine enthusiasm within your teams. This playbook is designed to guide you through both aspects, offering immediate practical steps alongside deeper strategic thinking.

The power of quick AI wins helps build momentum. In the initial stages, regularly securing quick wins—simpler AI chatbots delivering visible benefits in a short timeframe—is instrumental. They build enthusiasm by showing AI's tangible potential. This early positive energy is infectious. Quick wins also demonstrate value and secure buy-in, making it easier to justify further investment. For hesitant team members, experiencing a direct benefit can overcome skepticism, such as AI significantly reducing time on routine emails. Engaging with AI to achieve quick wins helps staff build basic skills and confidence. Examples of potential quick wins include using generative AI to brainstorm initial marketing campaign ideas (as discussed in Chapter 5 on prompting), assisting with first drafts of guest communications (Chapter 6), or creating unique visuals for social media (Chapter 7). These wins help manage expectations by showing immediate, although sometimes small, returns.

Although quick wins are crucial for initial traction, a string of isolated successes does not equate to a far-reaching AI strategy. To truly channel AI's power and achieve lasting competitive advantage, to become what we envision as an "AI Nexus Hotel" (explored in Chapter 15), your efforts must be guided by a clear, long-term strategic AI vision. This vision ensures AI initiatives align with core business goals, contributing to objectives like enhancing guest satisfaction, increasing RevPAR, improving operational efficiency, or building stronger brand loyalty. A strategic vision facilitates deeper, significant change by moving beyond task automation to reimagining core processes and business models, leading to more profound improvements. Furthermore, a long-term view helps make informed decisions about resource investment and encourages proactive planning for skills, data infrastructure, and ethical governance (Chapter 8), preparing your hotel's future. Appendix C provides a detailed phased roadmap that embodies this long-term strategic thinking, and understanding this balance upfront is key to how you approach the insights throughout this book.

The most effective approach is not to choose between quick wins and long-term strategy but to make them work together. Think of quick wins as fuel for your journey toward the larger strategic vision. Prioritize early AI use cases that offer immediate relief or value and align with one of your broader strategic themes. For instance, if a long-term goal is to hyper-personalize the guest experience, an early quick win might be using AI to help craft more customized welcome amenities. By understanding this balance, hoteliers can strategically build momentum, demonstrate value early, and keep their teams engaged while steadily progressing toward a more sophisticated and impactful AI-integrated future.

A Human-Centric Goal: Reducing Toil to Amplify Joy

A core principle guiding this playbook is the belief that AI's greatest potential in hospitality is to make work more fulfilling. To understand this, we must first distinguish between two types of work that fill every hotelier's day: toil and joy.

Toil represents the high-effort, low-reward tasks that are often necessary but drain our energy and time. This includes repetitive, administrative, or rule-based work that does not require our unique human creativity, empathy, or strategic judgment. Think of manually compiling data for daily reports, repeatedly drafting nearly identical confirmation emails, or spending hours searching for information across different, disconnected systems. While essential for operations, these tasks offer little professional satisfaction and can be a significant source of burnout.

Joy, in contrast, is the work that energizes us and leverages our best human abilities. This is where we create true value and find meaning in our profession. It involves creative problem-solving for a guest's unique situation, building genuine relationships that foster loyalty, strategically planning a new service offering, or mentoring a team member to help them grow. This is the work of genuine hospitality, where empathy, intuition, and personal connection make all the difference.

Understanding this distinction is fundamental to the AI adoption strategy presented in this book. The goal is strategically targeting and reducing the toil in our daily operations. By delegating these repetitive and draining tasks to AI assistants, we can dedicate more time, focus, and energy to the joy of creating exceptional guest experiences. This human-centric approach transforms AI from a simple productivity tool into a powerful partner for elevating the art of hospitality itself.

Why This Book, Why Now

I wrote this book because too many smart people are left out of AI conversations. No one has taught them how to think differently, build the right habits, or trust these tools as collaborators instead of threats. The Generative AI Productivity Paradox is not a technology problem but a literacy problem. And once you solve for literacy, everything else—efficiency, satisfaction, innovation, even joy—can follow. This book is the first step.

PART 1

FOUNDATIONS

1
WHAT IS THIS AI THING

Demystifying Artificial Intelligence for Hoteliers

ARTIFICIAL INTELLIGENCE IS reshaping how we run hotels, serve guests, and manage our time. We are talking about practical AI tools that can significantly change your operations, enhance guest experiences, and free up your team to focus on what truly matters.

Many hoteliers understandably worry if AI will diminish the crucial human touch that defines our industry. This book will show you how AI can free up your team to provide even more meaningful and personalized human interactions by handling the routine. If you are feeling curious or cautious about AI, that is normal. This technology represents real change, and it is change you can manage. Throughout this book, I argue that AI can be a powerful ally for hospitality only if we understand what we are working with. Real significant change requires a new perspective, starting with clarity about what AI is. This chapter lays that foundation, helping you build the AI literacy you need to confidently lead.

The Journey of AI: From Early Ideas to Today's Tools

Today's AI did not appear overnight. What feels like a sudden revolution results from nearly seventy-five years of development. The concept began in 1950 when British mathematician Alan Turing published a paper asking, "Can machines think?" He proposed the famous Turing Test to judge if a machine could demonstrate intelligent behavior through conversation. This question sparked decades of research. The term "artificial intelligence" was coined a few years later at a 1956 Dartmouth College workshop.

The first decades of AI research focused on symbolic AI or "good old-fashioned AI" (GOFAI). This approach tried to build intelligent systems by programming computers with explicit rules and logical processes. Scientists worked to encode human knowledge into databases and develop detailed if-then systems. These systems had some impressive successes in structured areas, like playing chess, but they struggled with the messier, detailed problems of the real world, especially understanding natural language or images. They could follow rules but had difficulty learning or easily adapting.

A major shift began in the 1980s and accelerated in the 2000s with the rise of machine learning (ML). Instead of programming every rule, ML uses algorithms that let computers learn from data. When fed large amounts of examples, these systems find patterns and improve over time without being explicitly reprogrammed. Within ML, artificial neural networks became indispensable. Inspired by the human brain, these systems use layers of connected nodes to process information. When these networks have many layers, it is called deep learning. Using massive datasets and powerful graphics processing units (GPUs), deep learning models achieved breakthroughs in previously challenging areas. Suddenly, AI could recognize objects in images, understand speech, translate languages, and power recommendation engines on sites like Netflix and Amazon. Even before the generative AI buzz, hospitality was already using these advances. Modern revenue management systems use complex algorithms to analyze past bookings, market demand, and competitor prices to optimize room rates. CRM platforms use ML for predictive personalization, suggesting offers to different guest segments based on behavior patterns.

The key technology for today's generative AI, which can create new content like text, images, audio, or code, arrived in 2017. Researchers at Google Brain introduced the Transformer model, which included a mechanism called "attention." This allowed AI to weigh the importance of different words in a text simultaneously, rather than processing them strictly in order. When you ask it to draft a marketing email about a specific package, it can pay more attention to the package details you provided, leading to a more focused and effective draft. This dramatically improved the ability to understand context and connections in language. Transformers proved highly efficient at learning from and generating language using enormous datasets. Research labs like OpenAI developed increasingly powerful large language models (LLMs) based on this transformer architecture. Their GPT (generative pre-trained transformer) series showed impressive ability to create fluent, coherent text. When ChatGPT launched in late 2022, its conversational interface and ability to generate human-like text on almost any topic captured worldwide attention, gaining over 100 million users in record time. This was a watershed moment: Generative AI moved from research labs directly into the hands of millions, ready to be applied across every industry, including hospitality.

We have access to a growing ecosystem of these powerful generative AI tools today. Beyond OpenAI's GPT models, there are other robust systems like Anthropic's Claude, known for safety and handling long texts, and Google's Gemini, designed for multimodal capabilities, and many specialized and open-source models. They are becoming practical, accessible tools for improving productivity, sparking creativity, and solving real business challenges in hotels.

How Does Generative AI Work?

So, how do these generative AI systems perform their impressive linguistic and creative feats? To use them effectively, it helps to understand their basic functioning without needing to be an engineer.

First, let's be clear: Current AI systems are without sentience or consciousness like humans. At their core, today's generative AI, especially large language models, are advanced pattern-matching and prediction engines. Think of it like a super-charged version of the auto-complete feature on your phone. If you type "Have a nice," your phone might suggest "day" because it has seen that pattern countless times. Generative AI works similarly but on a vastly larger and more intricate scale.

When you give an AI a prompt, your instruction or question, it analyzes that input using the complex web of linguistic patterns, grammatical structures, and factual associations it learned during training on billions of words. According to this knowledge, it predicts the most likely next "token" that should follow. A token can be a word, part of a word, or punctuation; think of these tokens as the AI's fundamental building blocks for language, like individual beads it strings together to form a coherent sentence or paragraph. It adds that predicted token, re-evaluates the new sequence, predicts the next most probable token, and continues building its response one token at a time.

These models do not "understand" your prompt or their generated text in a deep, human-like way. They have learned an incredibly complex statistical map of language: how words connect, in what order they typically appear, what writing styles are common, and what information is frequently associated with particular topics. The "attention mechanism" within transformers is crucial here. This helps the AI dynamically determine the importance of all the tokens in your prompt and everything it has written so far, which guides its prediction for the next token. This gives it an impressive ability to maintain context, follow instructions, and generate outputs that feel human-like, coherent, and sometimes surprisingly insightful. If an AI becomes skilled at predicting and assembling these token sequences in ways that match human language patterns, it can perform an impressive array of tasks, like drafting marketing emails, summarizing reports, brainstorming ideas, outlining training modules, translating text, and even generating code. Each of these is a structured sequence of linguistic or symbolic patterns.

Many generative AI tools also have a "temperature" setting. This lets you control the randomness or "creativity" in the AI's token selection process. A low temperature setting tells the AI to be conservative and predictable, usually choosing the statistically most likely token. This works better for tasks requiring accuracy, strict adherence to instructions, or logical precision, like summarizing legal documents or extracting data from reports. A higher temperature encourages the AI to explore less common, potentially more novel or unexpected, token choices. This helps with creative brainstorming or generating diverse ideas, although it slightly increases the risk of outputs being unfocused, inaccurate, or occasionally nonsensical.

Understanding this predictive nature of generative AI is important. It helps you move beyond seeing these tools as mysterious "black boxes" or as systems that think like humans. Instead, you can approach them as capable, understandable, and directable partners. This understanding is key for writing effective prompts, which we will cover in detail in Chapter 5. It also helps you evaluate the accuracy, relevance, and usefulness of AI outputs, a skill we will focus on in Chapter 9. Most importantly, it enables you to integrate AI into your hotel operations thoughtfully and responsibly, remembering that your human expertise, contextual understanding, and ethical judgment remain the ultimate guides.

The State of Generative AI: Real Results in Hospitality

A new 2024-2025 study by Vivander Advisors, sponsored by the HSMAI Foundation and AHLA Foundation, explores the impact of generative AI on commercial teams within the US hospitality sector. The research, titled "State of Generative AI in the Hotel Industry Commercial Teams," focuses on how these technologies are reshaping recruitment, retention, and professional development in sales, marketing, and revenue management. It highlights that the industry is at a pivotal crossroads, where traditional service excellence meets revolutionary AI capabilities, presenting both significant opportunities and challenges, especially for these revenue-generating functions. The study, led by Michael J. Goldrich, aims to provide actionable insights for organizations navigating this new technological environment.

The research reveals a fragmented and cautious approach to generative AI adoption across the hospitality industry. Most organizations are in early exploration phases, either testing AI with select teams or just beginning to consider its implementation. A significant enthusiasm gap often exists between optimistic leadership and more measured associates. Key implementation challenges identified include knowledge gaps, strategic uncertainty, and overall organizational readiness. The study indicates a prevalent absence of formal AI usage policies in many organizations, signaling an urgent need for structured governance as AI integration progresses.

Generative AI is fundamentally altering talent management strategies. Recruitment is shifting toward valuing candidates with hybrid skills, blending traditional hospitality knowledge with AI fluency and customer engagement capabilities. Traditional roles are evolving. This change necessitates new approaches to professional development, emphasizing continuous AI learning and critical thinking to complement established hospitality competencies.

The path forward for the hospitality industry involves a thoughtful, developing approach to AI integration, balancing technological innovation with the indispensable human touch. Success hinges on developing adaptive frameworks, investing in comprehensive AI literacy programs, establishing clear ethical oversight, and fostering transparent communication. Hotel technology partners are seen as crucial allies in this journey, helping to bridge the gap between current capabilities and future needs. Ultimately, organizations that strategically embrace AI as a tool to augment their workforce, reduce toil, and enhance meaningful work will be best positioned to define the future of hospitality.

Key Takeaways:

- AI possesses a rich history, evolving from rule-based systems, like symbolic AI, to data-driven learning methods, such as machine learning and deep learning; the transformer architecture is pivotal to today's potent generative AI.

- Generative AI functions as a sophisticated pattern-matcher and prediction engine; it forecasts the next most likely token based on its extensive training data and lacks true human-like understanding or consciousness.
- Recognizing this predictive mechanism helps you craft effective prompts, critically assess AI outputs, and utilize AI as an understandable, directable collaborator.
- AI's capacity to quickly generate text, summarize information, and assist with creative tasks can save considerable time and lessen the mundane aspects of work, allowing greater focus on strategic and guest-centric activities.

Bringing This into Focus:

- **Reflect on AI in Your Daily Life**: Consider where you already encounter AI, for instance, in recommendation engines on streaming services, predictive text on your phone, or smart assistants. How does understanding its predictive nature alter your perception of these tools?
- **Observe AI Outputs**: If you can access a generative AI tool, ask it a simple question or assign it a basic task. Examine its response. Can you perceive how it might be stringing together probable language patterns rather than thinking in a human way?
- **Discuss with a Colleague**: Share one key insight you learned from this chapter about what AI is with a colleague. Explaining it to someone else is an excellent method to deepen your understanding.

Leadership Action Agenda:

- **Demystify AI for Your Team**: At your next team meeting, allocate ten to fifteen minutes for a simple, non-technical explanation of what generative AI is, perhaps drawing from the prediction engine analogy in this chapter. Concentrate on demystifying it and reducing apprehension. Your aim is to start building a shared foundational understanding.

- **Host a "What is AI?" Q&A Session**: Offer a brief, informal Q&A session where team members can ask basic questions about AI in a safe setting. This helps gauge current understanding and address initial concerns.
- **Identify One Toil Task for AI Exploration**: Consider one highly repetitive, time-consuming administrative task in your department. According to what you learned in this chapter, consider: If an AI could reliably assist with just 20% of that task, what impact might that have on your team's time and focus?

2
CULTIVATING THE AI MINDSET

CHAPTER 1 EXPLORED artificial intelligence, its development, and how current generative AI can help reclaim time and significantly change hospitality. Understanding the technology, however, is insufficient. To move from occasional AI experimentation to strategic advantage, we must develop two vital human capabilities: AI literacy and an AI mindset. These are the foundations for successful, lasting AI integration in your hotel.

Still, these foundations can't develop in isolation; they need a supportive ecosystem. If the environment is unsuitable, even the best intentions for literacy and mindset development will falter, leading to wasted resources and missed opportunities. This indispensable ecosystem is built on the 4 Ts: tone from the top, tools, training, and time.

Understanding the 4 Ts

My belief in the 4 Ts framework comes from direct experience. Early in my work bringing AI to hospitality, my team delivered a one-day AI workshop for a hotel group. Initial feedback was enthusiastic, leaders were excited, and team members were energized. Adoption soared, and specific efficiencies were noted almost immediately in the pilot departments. Just kidding. That is what I hoped would happen. The reality? A year later, almost nothing had changed in their daily operations. Despite the initial enthusiasm, AI adoption remained minimal. Why? The organization had failed to provide specific, approved AI tools; there was an absence of ongoing training or dedicated time for practice, and although leadership had approved the workshop, a consistent, visible tone from the top promoting AI as a strategic necessity had failed to take hold. Without this supportive ecosystem, the seeds of literacy were unable to sprout. This was a valuable lesson: AI literacy is vital but effective only when supported by the right organizational structures.

When we worked with that hotel group again, we rebuilt the approach around the 4 Ts. The results this time were dramatically different, leading to tangible adoption and operational improvements. Let's examine why each pillar matters.

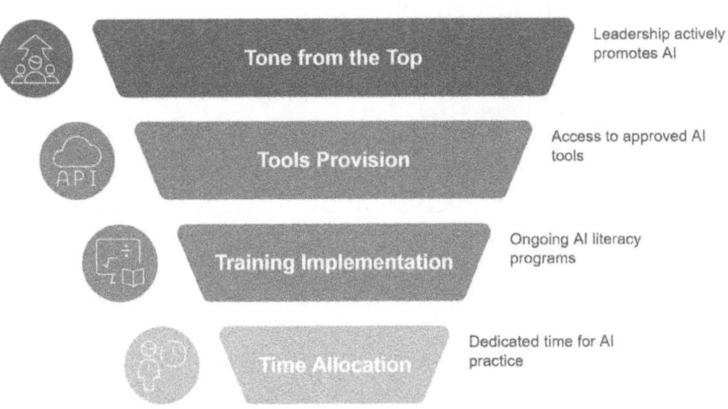

First is tone from the top. This goes beyond passive executive approval; it requires active, visible, consistent support. In our successful second project with the hotel group, the CEO did more than just approve the AI plan; he kicked off the training sessions. AI was presented as a core business need, directly linked to competitive advantage, operational efficiency, and guest satisfaction. This strong leadership signal, showing AI truly matters, was essential for organizational buy-in and reducing change resistance.

Next are the tools. If AI tools are hard to access or overly complex, or if there is confusion about approved platforms, widespread adoption will fail. In our first attempt, a clear tool strategy was absent; people experimented with free public AI, raising legitimate data privacy concerns. The second time, leadership made a definitive decision: Everyone in the pilot group received access to a specific, enterprise-grade AI platform. This was crucial. It meant consistent tools on a secure platform, addressing apprehensions about sensitive hotel and guest data.

Third is training, and this is where a dedicated program like the AI Literacy Mindset Program becomes essential. It is designed to be far more than a one-off workshop. It recognizes that learning to work effectively and thoughtfully with AI is a continuous, developing process. The program embodies the principle of genuine, ongoing, and role-relevant skill development.

Finally, the most precious and often hardest T to secure: time. If you expect busy teams to find extra hours to learn and use AI on top of their fully loaded schedules, you are setting them up for failure. Leadership in the successful project understood this reality. There was an open acknowledgment that short-term productivity might dip slightly before long-term benefits emerged. If teams struggle when first engaging with these new AI tools and adapting to new workflows, that sensation of challenge is often precisely what genuine learning feels like. Learning is work and requires dedicated time as a necessary investment toward achieving lasting AI proficiency and confidence.

The results from this 4-T-driven approach were unmistakable. After just six weeks, team members reported saving an extra hour daily. They produced better work, tackled more strategic projects, and enjoyed improved work-life balance. Daily AI usage dramatically increased. Participants consistently cited clear leadership (tone), good tools (tools), relevant learning (training), and protected time to practice (time) as the keys to their success. This experience showed that successful AI adoption involves more than finding one perfect app. It is about creating an ecosystem: a supportive environment where leadership, tools, training, and time work together. The training and time you invest are most effective when focused on building AI literacy and an adaptive AI mindset.

AI Literacy

To truly exploit AI's potential in your hotel, we must first master AI literacy. This is the bedrock of effective AI engagement. It means developing a clear, working grasp of AI: what it is, how it functions, its capabilities, and its limitations, all without needing to become an AI engineer. It fundamentally involves learning to interact with AI responsibly, ethically, and with a productive focus on your hotel's objectives. Think of AI literacy as achieving fluency in a vital new language—the language of human-AI partnership. While traditional literacy enables us to read and write text, AI literacy allows us to expertly read what AI produces, scrutinizing it with a critical eye, as detailed in Chapter 9. It also teaches us to write precise, context-rich prompts, which we explore in Chapter 5, that command AI to deliver valuable, on-brand results. AI literacy is your practical, empowering toolkit for working with conviction alongside this new intelligence. For hoteliers, AI literacy relates to developing a solid working understanding of several key areas.

- **Conceptual Clarity**: This involves understanding what you are working with when using AI. It signifies moving past buzzwords and hype toward a clear, practical view of what today's generative AI can and cannot do for your hotel. You should see these tools primarily as advanced pattern-matchers and content co-creators. This clarity also extends to knowing the four primary types of AI functionality impacting hotels:

- Creative AI: Your assistant for generating new content, like text or images
- Predictive AI: Your strategic engine for forecasting future events like guest demand
- Analytical AI: Your insight detective for finding hidden patterns in existing data, like guest reviews
- Automated AI: Your workhorse for executing repetitive, rule-based tasks

Understanding which AI type to apply is important, as mismatching the tool to the task is a common source of frustration. This overall clarity is key for setting realistic expectations and designing effective AI-assisted workflows.

- **Practical Interaction Skills**: This involves getting comfortable and effective using AI tools and is where understanding meets hands-on use. Can you and your team easily navigate a chatbot interface, detailed in Chapter 4? More importantly, can you take a hotel need, like responding to a guest email or brainstorming a promotion, and turn it into a clear prompt AI can understand? And then, can you evaluate the AI's output for tone, accuracy, relevance, and brand alignment?
- **Hospitality-Specific Application**: This involves making AI tools directly relevant to your hotel, guests, and team. It is one thing for AI to write a poem; it is more valuable to guide AI to draft an empathetic response to a negative review or help your F&B team find ways to reduce food waste, for example, by having AI analyze past spoilage data against purchasing patterns to suggest optimized ordering. AI literacy in hospitality means seeing and acting on

practical connections between AI tools and your daily tasks and strategic goals. It points to using AI's general abilities for specific hotel uses that save time and improve service.

- **Critical Evaluation and Responsible Use**: This involves using your human judgment and ethical compass, which is so important that we cover it in detail later in Chapter 8. Can you spot AI "hallucinations," which are false information presented as fact? Can you check the AI content for bias? Does everyone on your team understand and practice data privacy and information security, never inputting sensitive guest or hotel data into unapproved public AI tools? This ability to be responsible and ethically aware of AI is key.

The good news is foundational AI literacy, particularly the mechanics of using tools and basic prompting, can often develop relatively quickly with good, practical training using role-specific hospitality examples. Experience shows teams can move from AI novices to confident initial users in about four to eight hours of focused learning, much like learning a new software application. Still, a word of caution: This quick initial learning can sometimes create a false sense of mastery. Knowing how to use AI tools is only the first step. It is essential but insufficient if your goal is real, lasting change through strategic application. For that deeper, strategic level of AI integration, you must build the other foundation: a strong, adaptive AI mindset.

AI Mindset

If AI literacy is what you know about AI and how you use its tools, an AI mindset relates to how you and your team think about AI's role, strategic potential, and integration into your daily work and overall strategy. An AI mindset is less about specific technical skills and more about developing your mental models, problem-solving approaches, strategic foresight, critical evaluation abilities, and vision for human-AI collaboration in hospitality. This mindset separates those who use AI as just a faster tool from those who use it as a thinking partner and innovation catalyst.

A core tenet of developing this AI mindset is the commitment to daily, almost reflexive, consideration of AI's potential in every task. It means consistently asking, "How can AI help me with this?" and then actively attempting to apply an AI tool to that task. This daily practice is fundamental. It is perfectly acceptable, even expected, if AI fails to provide a perfect solution every time or if the initial attempts feel awkward or slow. Fueled by curiosity, this experimentation is a vital part of the learning process and a hallmark of a true AI mindset. These "failed" experiments are as valuable as successes, as they build deeper understanding and refine your approach.

Cultivating an AI mindset involves fostering key traits.

- **Strategic Curiosity and Visionary Thinking:** This pushes you beyond "Can AI draft this report faster?" to deeper questions, like "How could AI reshape our entire approach to operational reporting, making it a dynamic, predictive dashboard? Could this help us anticipate challenges or spot revenue opportunities before they happen?" It sees AI as a force for new service levels, operational agility, and innovation.
- **Relentless Problem-Solving Orientation:** This means seeing AI as a data-savvy partner in diagnosing root causes of ongoing challenges and co-designing smarter solutions. The focus shifts from "How can AI help our front office process guest emails faster?" to "How can AI analyze past inquiries to discover why we get so many recurring questions? And then, how can AI help us develop clearer pre-arrival information or better self-service options to reduce these routine inquiries, freeing our team for more complex interactions and saving everyone time?"

- **Embracing Iterative Collaboration with AI:** The most effective AI users rarely treat AI like a vending machine where one prompt goes in and a perfect answer comes out. Instead, they use a dynamic, conversational approach. They see AI's first output as a "version 1.0" that invites human refinement. These "AI collaborators" provide feedback, ask clarifying questions, offer more context, request revisions, and engage in multi-turn dialogue to guide AI toward an optimal result. It is a partnership.
- **Data Savviness**: An effective AI mindset includes a practical appreciation for data's key role. AI models learn from data. The quality, relevance, timeliness, accuracy, and potential bias in the data you provide or that the AI was trained on greatly affect the AI's output. Understanding "good data in, good data out" and its opposite, "garbage in, garbage out," is vital. For example, if guest feedback data entered into an AI is incomplete or inconsistently tagged, the AI's analysis of service trends will be unreliable.
- **Healthy Skepticism and Discerning Judgment**: Critical thinking is perhaps one of the most important traits of an AI mindset. It is the commitment to always question AI-generated content before acting on it. It means cross-referencing factual claims, checking for accuracy and bias, and avoiding blindly accepting AI's fluent outputs, especially for high-stakes decisions. It is knowing when AI is a great assistant and when human experience or ethics must override its suggestions. We focus on this in detail in Chapter 9.
- **Adaptability and Persistent Growth Orientation**: The AI environment is constantly changing. An effective AI mindset embraces this dynamism with curiosity and resilience. It means being flexible, willing to adjust strategies and workflows as AI develops, and ready to let go of outdated processes when AI offers a clearly better alternative. Most importantly, it means seeing the AI journey as a perpetual

process of continuous learning, iterative refinement, ongoing professional growth, and collaborative exploration. This commitment to lifelong learning is key.

Unlike foundational AI literacy, which can often be kick-started relatively quickly, a true AI mindset develops more slowly through cultural development. It requires consistent real-world experience using AI for meaningful hotel tasks. It needs dedicated time and safe spaces for teams to reflect, discuss what works, and share insights. And it needs visible support from leadership, along with strong peer-to-peer learning. Teaching AI tool mechanics might take a week, but it often takes months of daily practice, coaching, and reinforcement before teams genuinely start thinking differently with AI as a valued partner. That is when truly significant results begin.

Why Your Hotel Needs Both Literacy and a Developed Mindset

So, we have two powerful human foundations for AI success: AI literacy, the "know-how" to use AI tools effectively, and an AI mindset, the "think-how" to approach AI strategically and critically. What if an organization develops one and neglects the other? This is where many AI efforts falter. For AI to deliver its promise in hospitality, robust AI literacy and a developed AI mindset must work together.

Imagine your team has strong AI literacy. They can use AI tools well. Still, if their AI mindset has failed to develop, their use of these tools often remains shallow. They use revolutionary tools to do the same old things, just a bit faster. This is the "tool trap." Team members are skilled with tools but lack the vision to use them for breakthrough improvements.

Conversely, picture leadership with a brilliant AI mindset and a compelling vision for AI-driven change. Yet, if their frontline teams lack foundational AI literacy, if they are unable to effectively use the tools, write good prompts, or critically interpret AI outputs, that grand vision remains stuck in a "vision vacuum." Great ideas lack the practical ability to be executed. This leads to initial enthusiasm without real impact.

The impressive result happens only when robust AI literacy and a developed AI mindset are intentionally, concurrently, and synergistically developed and nurtured. This is the "synergy effect." AI literacy provides the "how-to;" the AI mindset provides the strategic "what if" and the discerning "why." When both are strong, your teams do more than just use AI more efficiently; they begin to discover and pioneer new, high-value AI applications. They move from executing pre-scripted AI tasks to confidently designing AI-augmented solutions. The organizational focus shifts from automating current processes to fundamentally reimagining them with AI as an indispensable partner.

This powerful synergy, however, does not materialize from understanding alone. It is forged through daily practice. The consistent application of AI skills, changing learned knowledge and new perspectives into ingrained work habits, makes this synergy impactful. The time to experiment, a crucial pillar of the 4 Ts framework, is precisely where these essential AI-driven habits are nurtured, allowing literacy and mindset to translate into tangible results. This human-centric approach, focusing on both what your people know about AI and how they think with AI, supported by an ecosystem of strong leadership tone, accessible tools, ongoing training, and dedicated time, is the key to unlocking AI's potential. It is how you significantly change your hotel into one that thrives with AI, fostering a culture where technology enhances human talent, reduces toil, and creates more time for the joy of exceptional hospitality.

Key Takeaways:

- Successful AI adoption requires a supportive ecosystem, known as the 4 Ts: tone from the top, appropriate tools, ongoing training, and dedicated time to experiment and learn.

- Two human foundations are indispensable for AI success: AI literacy, which involves understanding AI and how to use its tools effectively, responsibly, and critically, and an AI mindset, which means thinking strategically, curiously, and adaptively about AI's role.
- Developing a true AI mindset is accelerated by the daily habit of assessing every task with the question, "How can AI help me with this?" followed by active experimentation with AI tools; this consistent, curious engagement is key to integrating AI into your natural workflow.
- Neglecting either AI literacy or the AI mindset leads to suboptimal outcomes, such as the tool trap (good skills but no vision) or the vision vacuum (good vision but no skills); true significant change, the synergy effect, happens when both are developed together, enabling your hotel's AI ascension by saving time, reducing toil, and enhancing joy.

Bringing This into Focus:

- **Reflect on the 4 Ts in Your Hotel**: This week, honestly assess your hotel or department against each of the 4 Ts. Identify areas of strength and the most significant gaps that might hinder successful AI adoption.
- **Identify One AI Literacy Gap for Your Team**: Consider your team's current skills. Pinpoint one specific aspect of AI literacy, such as understanding hallucinations, basic prompting techniques, or ethical data use, where they might need immediate, focused learning.
- **Consider One AI Mindset Trait to Foster**: Determine which AI mindset characteristic, for example, strategic curiosity, iterative collaboration with AI, or healthy skepticism, would most benefit your team's current approach to

work or problem-solving, even before widespread AI use. Think about how you could begin to cultivate this trait.

Leadership Action Agenda:

- **Lead a 4 Ts Self-Assessment with Your Leadership Team**: Dedicate time in your next leadership meeting to discuss the 4 Ts. Ask each leader to rate your organization's current readiness on each T and to identify one concrete action the leadership team can take in the next thirty days to strengthen the weakest T.
- **Launch a Foundational AI Literacy & Mindset Kick-Off**: Plan a short, engaging kick-off session for all staff. The immediate goal is not deep technical training but to introduce the concepts of AI literacy and mindset, share the hotel's vision for AI as an empowering tool focusing on reducing toil and creating time for joy, and clearly articulate leadership's commitment, establishing the importance of AI adoption tone.
- **Champion an AI Learning Hour Pilot**: Select one department or team to pilot a weekly, protected AI Learning Hour for one month. Provide them with simple learning resources and encourage them to explore an approved AI tool. Gather feedback on what they learned and how the dedicated time affected their experience and skill development.

3
THE ROLE AUDIT

WE HAVE ESTABLISHED that cultivating AI literacy and an AI mindset supported by the 4 Ts is essential. You are likely now seeing the broader potential or perhaps wondering exactly where to start applying AI within your hotel's complex operations to get the best results. Simply gathering your team for a general "What can AI do for us?" brainstorming session, although often initially energizing, can result in a scattered list of ideas without a clear strategic path forward. We need a more systematic approach to using AI's power effectively and ensuring it helps reduce toil and create more joy. This chapter introduces a practical, step-by-step framework. It is designed to move you beyond general enthusiasm for AI and into a detailed understanding of your hotel's operations. The process involves carefully dissecting the work your team performs every day. From this, you can identify specific tasks and workflows where AI can demonstrably reduce burdensome effort, free up valuable time, enhance operational efficiency, improve decision-making, and elevate guest experience. This methodical approach will help you pinpoint high-impact AI use cases strategically aligned with your hotel's goals and contribute positively to your team's work life. To effectively track this value, it's important to establish key performance indicators (KPIs). A comprehensive list of suggested KPIs to measure AI implementation success can be found in Appendix C.

Start with the Work

The journey to effective AI adoption begins with a foundational step: achieving a crystal-clear understanding of what your team members do every day. Too often, I have seen hotel teams attempting to "bolt on" AI capabilities to the visible surface of their established roles and processes without first taking time to thoroughly unpack what those roles and processes truly comprise. And genuine significant change rarely occurs at this surface level. It emerges from a deeper understanding of the underlying mechanics of daily work, the intricate web of micro-tasks, frequently repeated actions, and often invisible workflows that consume your team's time, energy, and focus. That is why every successful AI implementation begins with something other than an outward look at the technology and with an inward look at the work itself.

The starting point is a candid inventory of actual work performed. It's so much more than a job description. It's the comprehensive collection of everything an individual or team does daily, weekly, monthly, and sometimes annually to keep your hotel running. From responding to guest inquiries and preparing sales proposals to participating in revenue strategy meetings and entering data into multiple systems, from crafting marketing content and generating financial reports to checking operational systems, coordinating between departments, and solving unexpected problems, this is the actual lived, day-to-day experience of work in your hotel.

Now, take that real job and break it down. This is the essence of job deconstruction. First, identify the major tasks or core areas of responsibility that define the role. For a director of sales, these might include developing quarterly strategic sales plans, managing key corporate accounts, leading and mentoring the sales team, and overseeing group booking conversions. Each is a significant chunk of their overall responsibility. Similarly, a room attendant's major task of cleaning a guest room to standard could be identified. Next, for each major task, drill down to more specific, discrete sub-tasks, those sequential steps typically required to complete that larger objective. For overseeing group booking conversions, relevant sub-tasks might include reviewing incoming requests for proposals (RFPs), assigning qualified leads to individual sales managers, monitoring proposal submission deadlines, and analyzing booking conversion rates by different market segments and lead sources. These are the direct actions that contribute to the major task. For our room attendant, sub-tasks for cleaning a guest room might include making the bed, cleaning the bathroom, dusting and vacuuming, and restocking amenities. Then, for sub-tasks that are particularly critical, time-consuming, or ripe for improvement, dare to go even deeper and break them down into sub-sub-tasks. These are the most granular, often almost invisible actions: the individual clicks to find information, the repetitive lookups in the PMS or CRM, the rewrites of standard email phrasing, the endless copy-pastes between systems that fail to integrate, the small rule-based decisions made throughout the day, the quick information searches, and the mental summaries and categorizations of data that happen constantly and are rarely documented as formal work. For the room attendant's task of restocking amenities, sub-sub-tasks might be checking par levels for soap via an app, notifying the supervisor of low linen counts via text, or physically transporting supplies from cart to room.

This microscopic level of work breakdown is precisely where the richest AI opportunities often live. At this sub-sub-task level, individuals can begin their AI adoption. By evaluating in such minute detail, team members start to see clear opportunities where AI can assist, freeing up personal time, reducing toil, and allowing focus on engaging aspects of their roles. In hospitality, most significant opportunities for AI are typically lacking at the broad task level. They are more often hidden within specific sub-tasks or seemingly insignificant sub-sub-tasks. It is rarely about AI handling all guest communication; it is more likely about AI intelligently assisting with sub-sub-tasks, like drafting an initial response to a common guest inquiry about late check-out availability or extracting client preferences and contact details from our CRM and populating them into our group sales proposal template. It is probably something other than AI analyzing all guest feedback autonomously; it is more likely about AI helping to accurately tag sentiment themes and specific service issues mentioned in thousands of guest review comments, making human analysis more efficient and insightful. These small micro-tasks, often dismissed as just part of the job, are frequently where significant amounts of your team's time are lost, where operational friction accumulates, and where employee morale can be drained by the daily grind of repetitive, unfulfilling toil work.

This structured decomposition of work is more than just an intellectual exercise. It is the bedrock of identifying opportunities for intelligent automation, AI augmentation, and designing effective, human-centric workflow improvements. The innovative team at SmarterX, the minds behind JobsGPT, has pioneered one of the most rigorous approaches to this challenge. Their analytical system is designed to break down complex jobs into atomic-level tasks and sub-tasks, then map each to a calibrated "AI exposure level." This exposure level indicates how likely that specific task is to be significantly enhanced by current and emerging AI models. The strategic insight is profound: You cannot discover meaningful, high-impact AI use cases for your operation unless you first take time to deeply understand the structure of your team's work. And you are unable to truly understand that structure unless you look beyond job titles or annual goals to see the operational seams, points of friction, repetitive loops, and inefficient workarounds. What specific actions do your team members perform repeatedly, almost on autopilot? What information are they constantly rewriting or reformatting? What tasks do they consistently avoid because they are tedious or time-consuming? What seemingly small activities add up to hours of lost productivity each week? That is precisely where AI often belongs first and where it can deliver the most immediate value by creating time and reducing toil.

When you and your team begin to see your work through this high-resolution lens of deconstructed tasks, something impressive happens: Profound clarity emerges. In all its multifaceted parts, the role becomes vividly visible, perhaps for the first time. With that newfound visibility comes genuine possibility. Once you truly understand your work and deeply understand its mechanics, pain points, toil components, and hidden inefficiencies, you can ask strategically: Where could artificial intelligence genuinely help us here? Could it accelerate this data extraction task, saving hours each week? Could it improve the quality and personalization of guest communication? Could it automate this repetitive sub-sub-task entirely, freeing up significant time for complex problem-solving? And perhaps most importantly, could it help create the precious space necessary for me and my team to dedicate more energy, passion, and unique talent to the parts of our jobs that truly matter—the parts that drive guest delight, foster team engagement, build value for our hotel, and, ultimately, bring us more professional joy? This thoughtful, introspective inquiry, starting with work deconstruction, is the beginning of your hotel's AI significant change journey. It starts with achieving self-awareness about the nature of the work itself. Only once you have carefully broken down your key roles and processes into their fundamental components can you strategically, intelligently rebuild them with artificial intelligence serving as your powerful, versatile operational engine.

Gauging AI's Potential: Understanding AI Exposure Levels

As you assess where AI can best fit within your hotel's operations, the conversation will shift from a general "What can AI conceptually do?" to a more strategic inquiry: "For any given role or task, how significantly exposed is it to what AI can already accomplish today, and what will AI likely be capable of doing very soon?"

The team at SmarterX offers one of the clearest frameworks for systematically answering this question. Their approach evaluates work at the granular and sub-task levels rather than the broad job title levels. They assess each using AI exposure levels rated on a calibrated scale (from E0 to E11 in their detailed modeling). This exposure level reflects how current and emerging AI models can significantly reduce the time needed to complete a specific task at or above the quality level expected from a proficient human colleague. A high AI exposure level for a task doesn't automatically signify that the task (or the valued human performing it) is slated for automation or replacement. However, a higher exposure level signifies greater opportunity. It is an opportunity for significant task acceleration through AI assistance, quality enhancement by leveraging AI's capabilities, intelligent delegation, or even fundamental redesign of how that task is performed to deliver greater value with less toil and in less time. This framework provides a lens for forecasting down to specific workflows and skill sets that must be developed.

We can conceptually group these exposure levels to better understand their implications:

An Absence of Meaningful Exposure (E0):

At the bottom of this exposure spectrum, we find E0, which describes tasks and interactions that currently have essentially an absence of meaningful AI exposure. These activities typically demand detailed in-the-moment human presence, sophisticated empathy, complex problem-solving in novel situations, or intricate physical dexterity. Think of the warmth and connection a skilled front desk agent conveys when welcoming a distressed guest who has just arrived after a difficult journey. Consider the subtle diplomacy and creative problem-solving required from a duty manager to resolve a sensitive guest service failure. These are, and will likely remain, the quintessentially human moments of hospitality. The human touch, guided by human wisdom and empathy, remains our industry's most precious asset.

Direct Task Augmentation and Enhancement (E1–E2):

Moving up the exposure spectrum, we encounter E1, which signifies direct AI exposure for immediate task enhancement. These are typically tasks where gaining access to a powerful large language model (like ChatGPT, Claude, or Gemini) can almost immediately reduce completion time and often increase output quality. Consider common tasks, like drafting a standard internal memo or guest email, summarizing meeting notes, or organizing brainstormed ideas. If you can paste the basic requirements into an AI prompt and quickly receive a helpful starting point, you are likely operating in E1 territory. This level offers tangible time savings and a boost in content creation capacity.

Next is E2, where AI augmentation comes from more than just general-purpose AI and increasingly from sophisticated AI-powered applications embedded directly within your existing software ecosystem. This is where platforms like Microsoft Copilot or Google's Gemini integration shine. These tools do more than just generate text in isolation; they often pull relevant context from your office suites to offer smarter, more contextual assistance. For example, an AI feature embedded in your sales team's Google Drive can automatically analyze client history and generate a customized follow-up email after a site tour, which is a clear E2 application. At this level, your team members are co-working with intelligent software designed for their existing digital environment.

Multimodal Interaction and Advanced Reasoning (E3–E8):

As we continue up the AI exposure scale, AI's capabilities expand in modality (E3–E6) and its ability to understand, process, and generate different types of information beyond text. The framework details these through E3 (image capabilities), E4 (video understanding), E5 (audio processing), and E6 (natural voice conversation). For a hotel, this means AI helps analyze guest-uploaded photos of room issues, generate concept art for a new restaurant theme, or power sophisticated voice-activated concierge services.

Further up, at E7, we encounter AI demonstrating advanced reasoning. Here, AI moves beyond content creation to provide genuine cognitive support. It can help solve multi-step problems, interpret ambiguous inputs, generate strategic options for complex decisions, and explain its reasoning. A revenue manager might prompt an E7-capable AI to analyze booking pace data, competitor pricing, and event calendars, then ask it to do more than just forecast occupancy. They might ask it to generate hypotheses for underperformance and recommend strategic actions. This is sophisticated cognitive amplification.

Then, at E8, the framework points to AI developing persuasion capabilities. At this level, AI can actively seek to influence behavior. These systems adjust language, tone, and arguments to shift preferences or guide choices. In hospitality, this might manifest in advanced AI-driven upselling agents on your booking engine or marketing chatbots that dynamically adjust messaging to maximize conversion.

Autonomous and Physical World Interaction (E9–E11):

By E9, AI is actively acting on your behalf within defined parameters in the field of increasingly autonomous digital agents (as explored in Chapter 13). These AI systems can independently perform task sequences or pursue broader goals, like processing group booking requests, checking PMS availability, drafting initial proposals, updating CRM, and scheduling sales manager follow-ups.

Finally, from E10 to E11, the framework looks at AI interacting with and acting within the physical field. At E10, AI paired with vision hardware (smart glasses, intelligent cameras) could "see" and respond to events in real time. At E11, AI becomes fully embodied, potentially through robots for specific, repetitive, or hazardous back-of-house roles like high-volume laundry, large-scale food prep, or inventory management. Although guest-facing embodied AI is further off, the technology is developing rapidly.

Why does this entire AI exposure level spectrum matter for you as a hospitality leader? Because virtually every task in your hotel is now a candidate for reassessment. Understanding a task's AI exposure level means that your team's precious time, unique talents, and invaluable human judgment can be strategically refocused toward higher value, more fulfilling, more creative, and more uniquely human contributions.

Uncovering Your Hotel's Unique AI Opportunities:

After carefully mapping your team's time and understanding how AI exposure levels might impact various tasks, the crucial next step is to translate that insight into a tangible, actionable AI initiative portfolio. This means collaboratively discovering where artificial intelligence can help your hotel right now, without some abstract future, and within your concrete operational realities. One of the most effective ways to do this is by facilitating a structured Use Case Identification Workshop with your key teams, ideally guided by a skilled facilitator, such as an AI champion or a leader trained in the methodology. This workshop is designed for something other than a formal lecture, where leadership dispenses AI concepts. It also shouldn't develop into an unstructured brainstorming free-for-all that generates a hundred exciting and ultimately unfeasible ideas. Instead, it is a carefully structured, interactive, and practical conversation. It invites each participant to thoroughly examine their current workflows, reflect honestly on tasks that drain their time and energy, and collectively explore where AI could realistically step in to save effort, improve quality, enhance guest experience, or unlock new efficiencies. This department-neutral method provides everyone with a shared methodology for discovering potential AI use cases grounded in their daily reality.

The core, guiding question for this workshop, the question that unlocks the most insightful conversations, is deceptively simple but profoundly revealing: Looking closely at all the essential tasks you and your team perform each week, what are the specific activities that consume too much administrative time or feel like pure toil?

To help spark deeper thinking and surface often-unspoken pain points, we typically initiate the workshop with a few targeted reflective prompts:

- "Consider your typical workday or workweek. What repetitive, almost mind-numbing tasks consume a disproportionate amount of your time and mental energy and fail to require much of your unique creative thought, strategic judgment, or empathy?"
- "What kinds of information or responses do you find yourself frequently rewriting or re-explaining multiple times a week, or even multiple times a day, to different guests, colleagues, or stakeholders?"
- "What specific types of information do you regularly spend time hunting down from various sources, manually formatting, or laboriously explaining to others?"
- "When do you find yourself engaged in the process of manually copying, pasting, or transferring information from one system or document to another, simply because those systems fail to seamlessly 'talk' to each other?"
- "What types of operational or administrative decisions do you make repeatedly throughout your day or week that are largely based on the same set of criteria or established business rules every time?"

By focusing the initial conversation on these common points of operational friction, these "energy vampires" that subtly deplete productivity and morale, we bypass the tendency to jump immediately to abstract AI strategy discussions. Instead, we tap directly into your team members' lived work experiences. You are asking for something other than grand, revolutionary AI ideas. With genuine curiosity and a desire to improve their work lives, you ask: "What specific tasks or processes consistently feel heavier, more cumbersome, more frustrating, or less fulfilling than they should be?"

This approach quickly surfaced prime candidates for AI augmentation in one Use Case Identification Workshop with a hotel's sales team. These included laborious rewriting of standard RFP email responses with minimal personalization, time-consuming manual completion of corporate proposal templates, frequently forgotten and crucial recapping of client calls and site visits, and the frustrating process of pulling lead data from various portals and re-entering it into their CRM. For a marketing team, similar questioning highlighted different pain points: constant reformatting of approved content for different email clients or distribution platforms, the creatively draining task of writing dozens of unique, keyword-rich meta descriptions for website pages, or the endless chore of resizing creative assets for different social media platforms. And in a hotel's operations department, this process might quickly pinpoint activities like daily manual collection and aggregation of guest feedback from multiple platforms or significant managerial time spent preparing operational summary reports for leadership meetings.

Notice a common thread? None of these tasks is particularly glamorous or earth-shattering in isolation. That is precisely the point. AI's initial, most tangible value often emerges in an absence of some grand, futuristic application and in these seemingly small corners of the typical workday and those countless essential micro-tasks where enhanced speed, improved precision, unwavering consistency, and automation of mind-numbing repetition can truly make a difference to your team's daily experience, their available time, and your hotel's overall efficiency. It is in these areas that delegating specific sub-tasks to a well-trained AI assistant can free up significant pockets of precious time and mental space for your human team, which can then be reinvested in more strategic, creative, guest-facing, and professionally fulfilling endeavors—the joy work that defines exceptional hospitality.

Connecting Discovered Work to Real Business Value

Once a team begins to identify potential AI opportunities within their workflows, a new kind of momentum takes hold. People across departments start seeing tasks everywhere that could potentially be delegated to or significantly accelerated with AI assistance. This initial surge of ideas is encouraging; it is a sign that the AI mindset, that culture of curiosity and continuous improvement, is beginning to take root. This foundational understanding is pivotal to your hotel's AI ascension, providing the launchpad to strategically advance compelling business cases into validated proofs of concept, which can then be effectively scaled across your operations. However, this enthusiasm also brings a new leadership challenge: the critical need for strategic prioritization. All identified tasks or potential AI applications will not be created equal in terms of strategic importance, potential impact on core business objectives, or feasibility for implementation. Every AI experiment, however well-intentioned, will not necessarily deliver meaningful or measurable value to your hotel in the short or medium term.

That is why the next step in this systematic discovery process is primarily something other than finding the "right AI tool" for every idea. Instead, it is a strategic one, requiring your leadership team to apply a robust filtering mechanism, asking tougher, business-focused questions about each potential AI use case: Which of these identified AI opportunities represent convenient, nice-to-have applications, perhaps saving a few minutes here or there for an individual team member? More importantly, which are truly consequential, holding clear potential to deliver significant, measurable, and strategically aligned value to the team, a key department's performance, or the hotel's overall business success and guest satisfaction?

To answer that vital strategic question effectively and consistently for each potential AI initiative, we need to look closely at the direct, tangible relationship between the work involved in performing a specific task (or improving a workflow) and the concrete value that task or workflow creates (or, conversely, the cost it incurs if done inefficiently, inconsistently, or poorly). Specifically, for each promising potential AI use case that has emerged from your team's deconstruction and brainstorming efforts, your leadership team should rigorously assess and attempt to quantify, even if just with informed estimates:

- **The Frequency and Volume of the Task:** How often is this task performed within your hotel? Is it a daily, hourly, weekly, or perhaps only monthly occurrence? How many different people or roles across your organization are involved in performing it? A task performed multiple times daily by many team members often presents a larger opportunity for positive AI impact (in terms of total time saved or consistency gained) than an infrequent task performed by only one or two individuals.
- **The Current Time and Effort Investment:** How long does this task take to complete using purely manual methods or existing tools? What is the estimated total human effort (perhaps measured in total staff hours or full-time equivalence for high-volume tasks) currently dedicated to this activity across the relevant timeframe (e.g., per day, per week, per month)?
- **The Downstream Impact and Critical Dependencies:** What happens immediately after completing this task in your current workflow? Are other processes or team members waiting for its timely and accurate output? What are the typical negative consequences—operational delays, increased errors, duplicated effort, guest service disruptions, or increased workload for others—if this specific task is performed slowly, inconsistently, or with errors, or isn't performed at all? Understanding these downstream dependencies is essential for gauging the true ripple effect of improving a task with AI.

- **The Strategic Importance and Direct Business Alignment:** Perhaps the most important filter for prioritization is answering this question: How does significantly improving this specific task through AI augmentation directly and demonstrably help your hotel move the needle on one or more core business objectives? For example, will this AI initiative tangibly improve a key guest satisfaction score? Will it contribute to enhanced staff retention, reduced employee burnout, or more effective new hire training? Will it reduce the time-to-decision for important revenue management strategies or key marketing campaign adjustments? Will it help your sales team respond faster, more intelligently, or more persuasively to valuable group booking inquiries or corporate RFPs? Or will it enable your operation to react with greater agility and insight to rapidly shifting market signals, developing guest preferences, or new competitive threats?

Let's consider two brief examples from a typical hotel to highlight the strategic difference between AI applications that primarily offer individual convenience versus those that deliver consequential organizational value. In our first scenario, imagine a sales coordinator who discovers they can use a generative AI tool to help them quickly rewrite and slightly personalize their standard thank-you notes sent after initial hotel site tours. It is a thoughtful application of AI, saving perhaps five to ten minutes of drafting time per lead. It is certainly a nice individual productivity boost. Still, the overall direct, measurable impact of this specific AI use on key hotel sales metrics, such as proposal conversion rates, total group revenue booked, or client retention, is likely modest. This is a perfectly acceptable, and likely lower-priority, convenient AI use case.

In our second scenario, picture a revenue analyst who, working with leadership, spearheads the development of a custom AI assistant (as detailed in Chapter 12). This system is designed to rapidly generate insightful initial demand summaries, flag unusual booking patterns in real-time, and draft narrative opportunity analyses directly from your hotel's raw PMS data and daily STR reports. This new AI-driven workflow shaves two to three significant hours off the previously manual weekly revenue strategy meeting preparation time for at least three key commercial team members. And, far more importantly than just time saved on report generation, this system consistently unlocks precious analytical capacity and cognitive bandwidth for your entire revenue team. They now find themselves able to dedicate far more meeting time to engaging in rigorous what-if scenario modeling, deeper collaborative strategic planning, and more proactive pricing and distribution adjustments. This delivers significant strategic value to the entire commercial function and the hotel's overall profitability. Both are valid AI uses. And only the second delivers the kind of significant organizational lift that justifies focused investment and strategic prioritization.

This is where the discipline of effective AI adoption truly begins: thoughtfully prioritizing AI use cases that create demonstrable, widespread, and strategically aligned organizational value, lacking isolated pockets of individual convenience. This often means asking for every proposed AI initiative: Does it tangibly improve the overall guest experience? Does it meaningfully support staff well-being or skill development by reducing toil and creating time for joy? Does it demonstrably reduce time-to-decision or improve decision quality for vital strategies? Does it genuinely help our hotel respond faster or more intelligently to market shifts? The best AI use cases often intersect with high task frequency, significant friction, and genuine strategic importance. When correctly identified and well implemented, they free up substantial time and energy, empowering your leaders and staff to focus on detailed judgment, creativity, exceptional service, and insightful strategic decisions.

Is Your Hotel's Data AI-Ready? A Critical Foundation for Success

As you begin to identify where AI can enhance your operations and build specialized tools like Custom GPTs or sophisticated automations, a key question must be addressed: Is our hotel's data ready for AI? The effectiveness of almost any AI system, especially for powerful applications like guest personalization, operational analytics, the knowledge base of custom GPTs, or the data quality needed for reliable automation, fundamentally depends on the quality, organization, and accessibility of the data it uses. Think of it as "good data in, better AI insights and performance out." Conversely, "garbage in, garbage out" is a stark reality. Before you can fully leverage AI to its potential, consider these key aspects of your hotel's data ecosystem:

- **Organized and Structured:** Is your data logically structured and consistently formatted? For AI to be effective, it must understand what it is looking at. This means guest preferences should be clearly tagged in your CRM, financial data should be uniformly categorized, and operational logs (like maintenance or housekeeping records) should be maintained with consistent terminology. Unorganized data is a significant barrier to effective AI.
- **Accessible:** Can your approved AI tools and platforms securely and efficiently access the necessary data streams? Data locked away in disconnected silos that AI is unable to reach is of little use for comprehensive analysis or intelligent automation. This often touches upon API availability and integration capabilities, which are vital for advanced AI functions.
- **Clean and Accurate:** Is your data free from significant errors, inconsistencies, duplications, and outdated information? AI models learn from the data they are fed. If that data is flawed, the AI's outputs, predictions, and actions

will also be flawed. For instance, multiple outdated entries for the same guest in your PMS will hinder any attempt at genuine personalization.

- **Relevant and Sufficient:** Do you possess the right type and sufficient volume of data for your intended AI applications? Meaningful guest behavior analytics require a rich history of interactions, and effective predictive maintenance for hotel equipment might require detailed operational data over time.

Investing time upfront to conduct a data readiness assessment is a strategic imperative. Clean, organized, accurate, and accessible data forms the bedrock upon which successful, reliable, and impactful AI solutions are built. Without this foundation, even the most sophisticated AI tools will struggle to deliver their promised value, particularly in areas like deep personalization and insightful analytics.

Understanding the Joy vs. Toil Balance

Our approach in deconstructing roles and identifying tasks for AI augmentation is essential for building a strategic foundation. Although critically important, time savings and efficiency gains represent only part of the complex equation that defines impactful and sustainable workplace change. Because even when two distinct tasks might take precisely the same amount of time to complete, they don't always feel the same to the person performing them. One task might spark engagement, flow, intellectual stimulation, or the deep satisfaction we call joy. The other might feel like a monotonous drag, a source of frustration, an activity that steadily depletes energy and motivation—pure toil. That is why, as we pinpoint AI opportunities, we must shift our lens slightly, moving beyond just the structure of work to also consider its sentiment and emotional resonance for our team members.

The Joy vs. Toil Audit, a concept created by researchers from Boston Consulting Group, picks up precisely where our more mechanical task mapping leaves off. It invites us to ask a human question about every activity we have identified: Of all the things you and your team do each day, which tasks genuinely give you energy, a sense of purpose, or creative fulfillment (the joy)? And which tasks consistently take energy away, feeling repetitive, draining, or disconnected from meaningful impact (the toil)? According to BCG's research, there appears to be a measurable threshold in this balance between joy and toil that significantly influences employee engagement, satisfaction, and even retention. Their data suggests that when employees spend more than four hours per week immersed in toil work, their likelihood of actively considering a job change or disengaging from their current role rises dramatically. Conversely, the research also found that if employees can consistently engage in at least ten hours of joy work each week, they are far more likely to report high job satisfaction, demonstrate greater loyalty, and contribute more proactively to the organization's success. Although a state of constant joy in every single task is an unrealistic workplace expectation for anyone, the data is increasingly clear: Sustained employee engagement, motivation, and retention depend significantly on leadership's ability to help tip this delicate scale decisively toward purposeful work and away from mind-numbing routine.

This Joy vs. Toil Audit is more than a feel-good productivity exercise. When integrated into your AI adoption strategy, it becomes a powerful diagnostic tool for gauging your organizational culture, assessing team morale, and identifying where AI can have the most positive human impact. An AI initiative focusing solely on raw efficiency gains will almost certainly fail to gain genuine traction and enthusiastic adoption from your team. Tools alone, no matter how intelligent, fail to transform teams. Relief does—relief from drudgery, relief from frustration, relief that creates space and time for more engaging and meaningful contribution.

The process of conducting a Joy vs. Toil Audit is, on the surface, deceptively simple. After carefully identifying and breaking down key roles into their core activities and tasks (using the job deconstruction method), you invite each team member to revisit their personal task list. This time, they are asked to label each recurring item by more than just its function, but by its feeling: Does this task primarily generate a sense of joy, or does it predominantly feel like toil? A joy task isn't always easy or fun in the conventional sense. It can certainly be challenging, demanding, and require intense focus. However, it is satisfying. It often aligns with an individual's core strengths, professional passions, or deeply held values. It might involve creative problem-solving, genuine human connection with a guest or colleague, strategic thinking that shapes future outcomes, mentoring a junior team member, or successfully navigating a complex operational challenge. These tasks often leave someone feeling energized and accomplished. Toil tasks, on the other hand, are typically energy draining. They often feel repetitive, reactive, highly manual, disconnected from a larger sense of purpose, or simply like a poor use of one's unique skills and talents. They might appear harmless: that daily report needing reformatting, those endless internal emails needing categorization, that data needing manual transfer. Over time, these accumulate, creating a slow bleed of morale and engagement.

By cross-referencing the identified emotional weight (joy or toil) of each task with the other critical data points you have gathered, a powerful, triage system for AI prioritization emerges. You are doing more than just identifying which tasks are technically automatable by AI. You are now identifying which automatable tasks also represent significant sources of toil for your team and, therefore, which AI interventions will most likely be embraced, appreciated, and deliver the greatest positive impact on employee well-being and engagement, alongside any operational efficiencies. You are pinpointing where AI can deliver to a more positive and sustainable work environment.

Your Blueprint for Strategic AI Implementation

This chapter has equipped you with a powerful set of interconnected frameworks for moving beyond abstract enthusiasm for AI and into the practical, strategic identification of high-value AI applications within your hotel. By starting with the work and applying job deconstruction, you gain an unprecedented, granular understanding of your daily operations. This foundational clarity allows you to accurately gauge tasks against AI exposure levels, identifying where current AI can offer the most significant assistance. The Strategic Use Case Identification Workshop, fueled by an honest assessment of operational pain points, becomes your engine for discovering AI opportunities that resonate with your team's real-world experiences. Connecting discovered work to real business value lets you prioritize initiatives that deliver consequential, measurable benefits. Finally, layering in the Joy vs. Toil Audit ensures that your AI strategy is human-centric, aiming to reduce drudgery and create space for more fulfilling, engaging work. Together, these tools form your blueprint for strategic AI implementation grounded in operational reality, focused on tangible value, and dedicated to empowering your team and delighting your guests. This approach moves AI from a conceptual possibility to a practical, significant force within your hotel.

Key Takeaways:

- **Job Deconstruction:** Break down roles into granular tasks (major tasks → sub-tasks → sub-sub-tasks) to see precisely where AI can help. This is foundational for AI ascension and identifying time-saving opportunities.
- **AI Exposure Levels:** Understand how susceptible different tasks are to AI augmentation or automation, using a framework like SmarterX's E0–E11 scale.

- **Use Case Identification Workshops:** A collaborative, pain-point-driven method to surface practical AI opportunities grounded in your team's daily toil.
- **Prioritize Use Cases:** Evaluate potential AI applications based on frequency/volume, current time/effort, downstream impact, and strategic business alignment, not just convenience.
- **Joy vs. Toil Audit:** Assess tasks based on their emotional impact on staff (drawing from BCG research), ensuring AI is targeted to reduce toil and enhance joy and well-being.

Bringing This into Focus:

- **Deconstruct One Task:** Choose one of your recurring work tasks. Try to break it down into at least three to five sub-tasks or sub-sub-tasks. Where does most of your time go? Which parts feel like toil?
- **Quick Joy-Toil Reflection for Your Team:** Briefly explain the joy-toil concept at your next team meeting. Ask each person to anonymously write down one task that feels like toil and one that brings them joy. Collect the tasks to see what patterns emerge.
- **Assess AI Exposure for a Deconstructed Task:** Take one sub-task you identified during your deconstruct one task exercise. Referring to the AI exposure levels, estimate the current AI exposure. Does this insight change how you view its potential for AI augmentation or the time that could be saved?

Leadership Action Agenda:

- **Sponsor a Departmental Job Deconstruction and Joy-Toil Audit:** Task one department head with leading their team through a job deconstruction for one or two key roles, followed by a Joy-Toil Audit for the identified tasks within the next month. Review the findings with them to identify high-priority areas for potential AI intervention that could save time and reduce toil.

- **Facilitate a Cross-Functional Use Case Identification Workshop:** Arrange a workshop focused on one specific hotel challenge (e.g., improving the pre-arrival communication experience or reducing F&B waste). Invite representatives from all involved departments. Use the pain-point questions from this chapter to generate five to ten potential AI use cases.
- **Initiate a Data Readiness Check for a Top Use Case:** Following the Use Case Identification Workshop, select one or two top-priority AI opportunities. Task the relevant department head, in collaboration with IT (if needed), to conduct a preliminary data readiness assessment for these specific use cases. Determine if the necessary data sources are organized, accessible, accurate, and of sufficient quality to support successful AI implementation for those high-value applications.

PART 2

ESSENTIAL AI IN PRACTICE

4
THE AI TOOLKIT

WE HAVE COME a long way. You understand what AI is. You've started to cultivate the right mindset, and you have frameworks for identifying where to focus your AI efforts. Now comes the practical question most hoteliers ask: "Which AI tools should I use?" This is where enthusiasm meets reality. With an ever-growing list of AI tools, how do you choose the ones that will help your hotel? What features matter? What is the difference between all these chatbots? And how do you avoid wasting time on shiny objects that fail to deliver results?

This chapter cuts through the noise. We will look at how modern AI interfaces work, explain the technical concepts you need to understand in plain English, and give you a framework for evaluating which platforms fit your hotel's specific needs. The goal is to make you a confident, informed buyer who can select tools that genuinely save time and drive value. (Fair warning: The AI environment develops quickly. Specific features, model names, and interfaces will change. As of this writing in summer 2025, the platforms mentioned are representative but always focus on the evaluation framework and selection principles; they will serve you well no matter what new tools emerge.)

The Anatomy of a Modern AI Chatbot Interface

Before choosing tools or mastering prompts, let's understand the environment where you will interact with AI daily: the chatbot interface. This is your control panel; understanding its components will make you more effective immediately. You will see a clean, minimalist design when you launch ChatGPT, Claude, or Gemini. Rest assured, underneath is a sophisticated system with several key components.

The conversation core is where your dialogue happens. This includes the message input field, your command center at the bottom of the screen, where you type prompts or your instructions to the AI. Most support multi-line entry using Shift + Enter, which is helpful for detailed prompts. The send button, often a paper airplane icon, submits your prompt. The chat history window is the central transcript showing your conversation history. This is vital because the AI uses prior conversation as context for new responses, up to its context window limit, which we will explain shortly. Session management options allow you to start new conversations and name or save threads, keeping your AI projects organized.

Beyond text, today's AI tools are increasingly multisensory; they can "see" by analyzing images, "hear" through voice input, and "talk" using voice output. Look for file uploads, often indicated by a paperclip icon. You can upload PDFs, such as meeting agendas to summarize action items or event BEOs to extract AV needs, Word documents, Excel spreadsheets to ask for data trend analysis or chart creation, or images to request descriptions or identify elements. Once uploaded, you can reference these in prompts, like "Summarize action items from this PDF" or "What trends do you see in this occupancy data?" Many interfaces also feature voice capabilities, letting you speak your prompts and have responses read aloud, perfect for hands-free use when you are prepping for an event or reviewing while on the move. Furthermore, these advanced voice features often support natural, multi-turn conversations, and many AI platforms offer a diverse range of voice options, including various languages, accents, and perceived genders, allowing for more personalized and inclusive interactions.

Within settings, you will find controls that significantly impact AI performance. Model selection allows you to choose between faster, economical models for simple tasks versus more powerful ones for complex reasoning. Memory settings help manage what the AI "remembers" across sessions, allowing personalization. Furthermore, many platforms are now incorporating AI connectors. These features enable the chatbot to securely link with your other essential digital tools, such as your email client, calendar, or even specific company knowledge repositories. By establishing these connections, the AI can access and utilize your personal or organizational information to provide more contextually relevant, personalized, and effective assistance for your tasks. Finally, many tools offer search capabilities that connect to the live web, overcoming the AI's knowledge cutoff date for current information, which is crucial for time-sensitive tasks like local events or weather forecasts.

It is most important to understand privacy and security controls. Reputable providers offer settings to manage data use. For business use, opt for enterprise platforms that guarantee your data will not be used to train public models. The golden rule is to avoid inputting guest PII, payment data, sensitive employee records, or proprietary hotel information into public, unapproved AI tools. Use only sanctioned, privacy-hardened corporate AI workspaces.

Key Technical Concepts for Hoteliers

Understanding a few basic technical ideas will make you a more effective AI user.

- **Tokens and Context Windows**: AI's "Working Memory"

  AI breaks text into tokens, which can be words, parts of words, or punctuation. Each AI has a context window, its short-term memory for a specific conversation. Early AIs had small windows, perhaps a few paragraphs, but today's advanced models can handle hundreds of pages. This matters because a small context window might cause the AI to "forget" early details for complex tasks like summarizing a lengthy report. For example, if you ask an AI to summarize a very long annual report for your hotel owners, a model with a larger context window is less likely to forget key details from the beginning of the document by the time it reaches the end. Knowing this helps you choose the right model or break large tasks into manageable chunks.
- **Training Data and Knowledge Cutoffs**: What Your AI Knows (and Doesn't Know)

  AI models learn from training data, which consists of billions of words from the internet, books, and other sources. This shapes their language skills, general knowledge, and potential biases. Generally, AI knows language patterns, grammar, writing styles, and broad factual topics, including general hospitality concepts. However, AI generally lacks knowledge of information after its knowledge cutoff date, your hotel's specific information, like exact room rates or a new chef's name, hyper-local knowledge not frequently documented online, and confidential company data. For example, if AI drafts website copy about your city, it will do well. But to have it highlight your brand-new rooftop pool, you need to provide those details. This limitation underscores the importance of providing detailed, current information within your prompts or leveraging Custom GPTs that can be trained on your hotel's specific, up-to-date knowledge base (a technique explored in Chapter 12) for relevant outputs.

- **Multimodal Capabilities**: When AI Gets Multisensory

  Multimodal means AI can process multiple formats: text, video, images, and audio. This makes it more versatile, engaging with information more like humans do. This is important because you can upload a dish photo, ask for a menu description, describe your lobby atmosphere, and get a visual concept. This ability to "think" across formats opens vast creative possibilities, saving time on content creation and basic analysis.

Understanding these concepts helps you craft smarter prompts, anticipate responses, and choose the right tools for specific tasks.

Selecting the Right AI Tools: A Discerning Look at Leading Players

The AI market is dynamic. As you explore the capabilities of various AI tools, it is natural to wonder about the investment involved. Many tools provide freemium models: a free basic tier, though often with limitations on usage or features. You will typically encounter monthly or annual subscription tiers for regular use, often based on features, capacity, or user numbers. Enterprise costs will likely be necessary for larger hotels or those needing advanced security, custom integrations, or higher usage volumes. These typically require a more significant investment and often come with essential features for governance and scalability. While this playbook avoids specific pricing due to rapid changes, factoring these potential costs into your planning is important for sustainable AI initiatives.

Expect to budget around $25 per person monthly for some leading platforms. For a hotel with 100 staff members, that is an additional $2,500 monthly, or $30,000 annually, for something likely absent from last year's budget. This presents a challenge: How do we value an investment that doesn't always fit traditional ROI calculations? Looking for a direct, easily measured ROI for this type of technology can be misleading. I explored this challenge in my master's thesis on the productivity paradox: the disconnect between technological investment and immediate, quantifiable productivity gains. Generative AI presents a similar dynamic, what I have termed the Generative AI Productivity Paradox, where potential meets underwhelming impact if incorrectly framed.

Instead of a pure ROI calculation, I urge you to look at AI investment through the lens of opportunity cost. What is the cost of *failing* to empower your team with tools that can give them back valuable time? For example, a 100-employee hotel group found individuals saved, on average, one hour per day using AI. Conservatively estimating a $20 average hourly wage, one hour per person daily translates to significant potential. For 100 staff, that is 100 hours saved daily, a conceptual $2,000 per day, and $480,000 per year in time value. The conceptual time ROI based on these figures is 1,600%. The true return is what your team *does* with that reclaimed hour. Emerging data shows people use this newfound time to accomplish more work, improve work quality, and devote time to strategic work. Although the $25 per month per person is a real cost, the opportunity cost of failing to enable that saved hour could be far greater. This reframing helps bridge the Generative AI Productivity Paradox.

Leading AI Platforms

Let's explore some leading AI chatbots:

- **ChatGPT (from OpenAI)**:

  ChatGPT has become almost synonymous with AI chatbot due to its conversational fluency, versatility, and an expanding ecosystem of capabilities, like the GPTs explored in Chapter 12. Its strategic partnership with Microsoft means OpenAI's models often drive Microsoft Copilot features. Powered by OpenAI's advanced LLMs (e.g., GPT-4 series), ChatGPT excels at creative content generation like marketing copy, blog posts, and social media updates. Advanced models show strong logical problem-solving and data analysis, especially via "Advanced Data Analysis."

  - *Key Hospitality Sweet Spots*: Hoteliers find ChatGPT invaluable for rapidly drafting personalized guest emails, generating creative taglines, outlining onboarding modules, or brainstorming F&B specials. Its ability to adopt a specific brand voice with clear prompting (Chapter 5) is a significant advantage. Integrated image generation via DALL·E and basic data visualization are useful for mood boards or charting KPIs.
  - *Important Considerations*: Accessing advanced models usually requires a paid subscription (ChatGPT Plus, Team, or Enterprise). Only paid versions like Team or Enterprise, which offer vital data privacy commitments and enhanced security, should be used for sensitive hotel or guest data. Public/free versions should be strictly avoided for confidential work due to privacy risks (a core tenet of AI governance, Chapter 8).

- **Claude (from Anthropic):**

  Anthropic emphasizes AI safety and constitutional AI, systems designed for helpfulness, harmlessness, and honesty. This is reflected in Claude's interaction style; its strengths include detailed reasoning, processing long documents, and a tendency toward caution.

  - *Key Hospitality Sweet Spots*: Claude's proficiency with extensive content makes it ideal for analyzing vendor agreements, comparing SOPs, summarizing training manuals, or drafting internal policies. A key differentiator for models like the Claude 3 family is their significantly larger context window, handling novel-length documents in one session. Users also praise Claude's ability to explain its reasoning. Anthropic's policy of avoiding the use of commercial API data to train general models provides confidence for confidential materials.
  - *Important Considerations*: Anthropic is actively working to enhance Claude's connectivity through initiatives like the Model Context Protocol (MCP), an open standard designed to help AI models like Claude integrate more seamlessly with various tools and data sources. Its third-party integration ecosystem may require more custom API work for some specific hotel systems compared to platforms with longer-established, deeply embedded marketplaces. Regarding multimedia capabilities, Claude models do not offer native image generation as of summer 2025.

- **Gemini (from Google):**

  Google's Gemini models stand out for inherent connectedness, native multimodality (text, image, audio, video, code), and real-time information access. Developed by

Google DeepMind and Google Research, Gemini integrates seamlessly within Google's ecosystem (Gmail, Docs, Sheets, etc.)

- *Key Hospitality Sweet Spots*: Gemini's direct access to live web information via Google Search is a major advantage for current information (flight delays, local events). Its native multimodality is powerful: Concierge can use it for local recommendations; marketing can analyze social media photos or generate campaign ideas according to travel trends; sales can generate talking points based on a client's latest news. Within Google Workspace, it assists in drafting emails, analyzing data in Sheets, or creating presentation outlines.
- *Important Considerations*: User experience can vary by access method (web app, Workspace integration, APIs). Access to advanced capabilities often requires premium subscriptions (Google One AI Premium or specific Workspace plans). Gemini's output can sometimes be verbose, potentially requiring precise prompting.

- **Microsoft Copilot**:

For teams in the Microsoft ecosystem, AI increasingly appears as Microsoft Copilot throughout Microsoft 365 tools (Word, Excel, PowerPoint, Outlook, Teams). It aims to understand work in context, significantly changing document creation, data analysis, and collaboration.

  - *Key Hospitality Sweet Spots*: Copilot's strength lies in the context of organizational data within Microsoft Graph and applications. In Excel, it helps analyze trends or generate charts. In Word, it helps rewrite or summarize. It summarizes email threads or suggests replies in Outlook. A revenue analyst can prompt Copilot in Excel for a forecast model, then generate a summary slide in PowerPoint. A front office trainer can draft an SOP in Word from a bulleted outline.

- *Important Considerations*: Effectiveness is maximized in a well-structured Microsoft 365 environment (SharePoint/OneDrive, Teams). Copilot's most comprehensive and feature-rich versions are generally available with enterprise-level Microsoft 365 subscriptions.

- **Grok (from xAI)**:

  Grok, from Elon Musk's xAI, is intentionally different, often exhibiting a bolder personality. Its key differentiator is real-time access to X (formerly Twitter) data, giving it an immediate pulse on current conversations and trends.

  - *Key Hospitality Sweet Spots*: Grok can be valuable for agile marketing, social media, and PR, needing to tap into developing social trends or public sentiment. A boutique resort might use it to surface trending hashtags. A brand manager could analyze X platform chatter around a local event.
  - *Important Considerations*: Grok's distinctive tone may be unsuitable for all brands (especially luxury), requiring careful prompt engineering (Chapter 5) and output filtering. Its broader knowledge or complex reasoning might be less mature for enterprise applications compared to leading models.

- **Open-Source Models (e.g., Llama family, Mistral AI models, DeepSeek)—Customizable Powerhouses:**

  Beyond proprietary platforms, an ecosystem of open-source AI models exists (e.g., Meta's Llama, France's Mistral AI), whose architecture and often pre-trained weights are publicly available. This allows organizations with technical capabilities to download, modify, fine-tune, and deploy these on their own infrastructure.

- *Key Hospitality Sweet Spots and Strategic Advantages*: This approach offers several strategic advantages for hospitality. Open-source models provide maximum data sovereignty and security, as self-hosting ensures that proprietary hotel and sensitive guest data never leaves your secure IT environment, which is essential for stringent compliance needs. They also allow for deep customization and fine-tuning on your hotel's specific datasets, such as training a model on thousands of guest reviews to perform detailed sentiment analysis that understands your property's unique context. While the initial setup requires investment, these models offer the potential for long-term cost savings for high-volume or specialized applications when compared to the recurring costs of commercial API pricing. Finally, they provide access to continuous innovation and community support, allowing skilled technical teams to leverage improvements and new techniques from the global open-source AI community.

- **DeepSeek (from DeepSeek AI)**:

Emerging as a significant global player, DeepSeek is gaining attention for coding, advanced reasoning, robust multilingual support, and computational efficiency, sometimes using a mixture of experts (MoE) architectures. Some models transparently show "chain-of-thought" reasoning.

  - *Key Hospitality Sweet Spots (Current and Potential)*: Strengths in many languages could support multilingual guest communication. Advanced reasoning can help with complex data analysis or operational process optimization. Their commitment to open-sourcing some models offers pathways for highly customized AI assistants.

- *Important Considerations*: Deploying open-source models requires significant in-house technical expertise. Capabilities in generating brand-specific marketing prose may be developing; rigorous evaluation is essential. Due diligence on data privacy is necessary. The user interface may feel more geared toward technical users.

Evaluation Criteria

Use these lenses to assess any AI platform.

- **Context Window:** Does this platform effectively manage extensive documents? For intricate assignments, like creating a detailed event proposal, a system with substantial memory is indispensable to process vast inputs without constant re-briefing, which improves final output and saves considerable effort.
- **Reasoning Capability:** Evaluate the model's cognitive depth. Simple text generation might suffice for standard emails; however, sophisticated thinking is required when tackling challenges, such as understanding declining elite-tier customer sentiment and formulating data-backed solutions.
- **Speed and Responsiveness:** Assess its performance for operational demands. Quickness is vital in frontline roles or time-sensitive situations where immediate replies are necessary for superior service. A sluggish application creates workflow bottlenecks and staff frustration.
- **Multimodal Capabilities:** Can the technology interpret various formats? A versatile architecture that comprehends words and images offers greater utility across departments.

For instance, marketing can conceptualize spa package descriptions with corresponding illustrations, while facilities could review pictures of damaged equipment to diagnose issues.

- **Data Protection and Security:** Verify the provider's commitment to information safeguarding. It is non-negotiable that your hotel's proprietary content avoids training external AIs. Confirm where your information is stored, what encryption standards are upheld (like SOC 2 or ISO 27001), and that your agreement explicitly prohibits your material from being used in generalized system development. Choosing a less-protected option presents a catastrophic liability.

Strategic Collaboration Beyond Single Tools

As your hotel's AI journey matures, the highest leverage often comes from strategically coordinating multiple AI capabilities, not relying on a single assistant. Different AI models possess unique strengths. Thoughtfully combining output unlocks new dimensions of accuracy, creativity, and insight. Though requiring more detailed management, this advanced approach can significantly enhance the quality and strategic alignment of AI-assisted work.

- **Cross-Verification and Collaborative Intelligence**:

  Cross-verification uses a second, distinct AI model to review, critique, or validate the output from a primary AI model, enhancing accuracy and reducing errors or biases. Instead of accepting a single AI model's output as definitive, especially for high-stakes decisions, use another AI for critique.

  - *Hospitality Example*: Your revenue strategy custom GPT predicts a 12% Q3 booking increase. Feed its assumptions into Claude for a critique of the logic. Ask Gemini to research emerging local market factors. This multifaceted approach promotes more robust decision-making.

- **Collaborative Prompting and Task Specialization**:
  This involves strategically assigning different project components to various AI models according to their specific "superpowers." Different AI platforms are better suited for different tasks.

  - *Hospitality Example (New Destination Wellness Retreat package)*: Marketing uses Gemini for travel trend analysis. F&B uses ChatGPT to brainstorm menu concepts and descriptions. Finance uses Microsoft Copilot in Excel for pricing scenarios. This mirrors a high-functioning human project team.

- **Sequential Prompting**:
  Sequential prompting uses one AI model's output as direct input for a subsequent model, creating an "assembly line" where each AI builds upon or refines previous work. This is an intelligent assembly line, with each AI adding a layer of refinement.

  - *Hospitality Example*: Ask Claude to summarize guest review themes about new spa facilities. Feed that summary to ChatGPT to draft a targeted email campaign. Provide the draft campaign to Gemini to review for cultural tone for international audiences. This pipeline can reduce errors and spark creativity.

- **AI Output Comparison**:
  This involves running the same prompt through multiple AI models and evaluating their varied responses to

identify the most effective approaches or insights. Avoid limiting yourself to one AI's first idea when seeking the most impactful message. Run the same prompt through multiple models and compare. This offers different stylistic approaches shaped by each model's architecture.

 - ○ *Hospitality Example Prompt*: "Write three compelling subject line options for a promotional email announcing a new luxury staycation package. Target audience: professional millennial women in New York City. Tone: chic, aspirational, inviting." ChatGPT might offer: "Your Ultimate NYC Staycation: Luxury Awaits." Claude might suggest: "Rediscover New York: An Exclusive Staycation Experience for the Discerning Woman." Gemini could propose: "Escape Without Leaving: Your Chic and Luxurious NYC Getaway is Here." You now have a richer selection of options.

- **Why This Multi-AI Approach Matters for Your Hotel:** Engaging with multiple AI chatbots and models encourages critical thinking, reduces over-reliance on any single system, builds broader AI literacy, and models a strategic mindset, showing AI as a diverse panel of specialized "advisors." This develops your AI tools from responsive assistants to proactive, integrated systems, turning a collection of tools into a strategic advantage.

Your AI Toolkit

By mastering interfaces, grasping key concepts, applying evaluation criteria, and envisioning integrated AI systems, you significantly change AI from a novelty into an indispensable capability. This mastery is key to reclaiming time and elevating performance and job satisfaction. Becoming comfortable navigating AI chatbot interfaces and grasping key technical concepts is your foundational launchpad.

Key Takeaways:

- Thinking beyond single tools toward coordinating a diverse AI ecosystem using techniques like cross-verification, task specialization, sequential prompting, and AI output comparison can significantly enhance strategic insights.
- Successfully leveraging AI involves framing the investment not just by direct ROI but by the opportunity cost of *not* empowering your team with tools that can save significant time and enable higher-value work, thus bridging the Generative AI Productivity Paradox.

Bringing This into Focus:

- **Explore an AI Interface This Week**: If you have access to ChatGPT, Claude, Gemini, or a similar tool, spend fifteen minutes exploring its interface according to the "Anatomy" section of this chapter. Try to identify all the components described. If the feature is available, attempt to upload a non-sensitive file to understand its functionality.
- **Test a Knowledge Cutoff**: Ask an AI tool about a recent event that happened yesterday or today. Then, ask about a well-known historical event. Observe the difference in how it might answer or if it indicates a need to use web search capabilities. This helps illustrate the knowledge cutoff concept in a practical way.
- **Consider One Evaluation Criterion for a Tool**: Think of one AI tool your hotel is considering or already using. From the five key evaluation criteria discussed (context window, reasoning, speed, multimodality, privacy/security), identify which criterion is most critical for that specific tool to be

successful in your hotel's environment and articulate why it is so important.

Leadership Action Agenda:

- **Facilitate a Toolkit Exploration Session**: Dedicate a portion of an upcoming team meeting to a hands-on exploration of one approved AI tool, using only non-sensitive data for practice. Guide your team through its interface, pointing out key features and settings. Encourage them to try any available conversation starters to understand the tool's range.
- **Discuss AI Tool Evaluation Criteria with Key Stakeholders**: Convene your IT lead, department heads, and perhaps your AI governance committee to review the key evaluation criteria outlined in this chapter. For your hotel's specific needs and strategic priorities, facilitate a discussion to rank these criteria in order of importance for selecting future AI tools. This creates a shared understanding for procurement and deployment, saving time on mismatched tools later.
- **Pilot a Multi-Tool Brainstorming Exercise**: For a specific, current challenge or opportunity your hotel is facing (e.g., generating new ideas for a weekend package), ask two different team members or small groups to use two different approved AI tools to brainstorm solutions. Have them present their top two to three AI-assisted ideas and discuss how the outputs differed. This exercise introduces the practical benefits of leveraging different AI strengths for creative problem-solving.

5
MASTERING THE PROMPT

WE HAVE EXPLORED AI's basic concepts, focused on AI literacy and mindset, and looked at the AI toolkit available to you. You now understand the why, the what, and the where of using AI in your hotel. Now, we come to the pivotal how, the skill that truly unlocks the significant potential of these intelligent systems: the art and science of prompting. This skill is central to your AI ascension, enabling you to directly influence AI to save time and shift your work from toil to joy.

To effectively collaborate with artificial intelligence as a digital coworker, a well-crafted prompt is your essential means of communication, allowing you to articulate your needs, provide context, and define desired outcomes. As with any valued human colleague, the usefulness of the AI's response hinges directly on the clarity, thoughtfulness, and strategic intent of your initial request. This chapter is dedicated to helping you master that vital interaction. This involves something other than learning complex programming. Effective AI prompting, especially with today's generative models, is mainly about clear human communication, sharp strategic thinking, creative direction, and understanding how these AI models "listen" and respond. It relates to developing your interactions from simple queries to sophisticated instructions that guide your AI partner in content creation, data analysis, operational problem-solving, and even strategic foresight. Mastering AI prompting is arguably the single most important practical competency for effectively leveraging AI. It is the difference between producing generic, unhelpful output and co-creating genuinely valuable, strategically aligned assets that save time, enhance guest experiences, and spark new insights.

The Fundamental Mindset

Before examining specific techniques, let's establish a fundamental mindset for interacting with AI, especially conversational chatbots. Imagine your AI assistant is a brand new, highly capable, incredibly intelligent remote associate who has just joined your team. It is their first day on the job. This associate is eager to help, can learn impressively fast, and has access to a vast amount of general knowledge. Still, like any new team member, they know nothing about your specific hotel, unique processes, established brand voice, current priorities, or the subtleties of a particular guest situation unless you tell them.

What do you do when onboarding a new human employee and assigning them their first significant project? You provide context by explaining the project's background and importance and how it fits into the bigger picture. You give clear instructions, defining the task, the specific deliverables, and the desired outcomes. You set expectations, outlining quality standards, timelines, and any critical parameters or constraints. You offer examples, perhaps showing them samples of previous successful work or templates to guide their approach. You establish guidelines, explaining relevant company policies, brand voice requirements, or ethical considerations. And when they deliver their initial work, you provide feedback, review their output, highlight what is good, suggest improvement areas, and offer guidance for refinement. This is exactly how you should approach prompting your AI.

Don't think of AI as an inanimate thing or an extraordinary black box; instead, it is a new, intelligent, remote associate. You are the manager, the director, the briefer. Your prompts are your instructions, context-setting, and expectation management. Your iterative refinements are your feedback. This mindset shift, from passively querying a tool to actively managing and guiding an intelligent collaborator, is the absolute key to unlocking consistently superior results. The more effectively you manage your AI associate with clear, comprehensive prompts, the more valuable and time-saving its contributions will be.

Two Fundamental Modes of AI Interaction

With this new remote associate mindset, let's revisit the two primary modes of interacting with AI.

First, there is conversational prompting. This often feels like an informal chat. This is like a quick, exploratory conversation for our new remote associate. You might casually ask: "Guest concierge (our conceptual guest relations AI), can you give me some quick ideas for handling a guest who is mildly upset about the pool being closed for an hour for unexpected maintenance?" This style is useful for initial brainstorming or getting a quick take. Still, just as you would not expect a detailed strategic plan from a new employee after a brief, vague chat, the AI's output here might be general. It is a good start, but its first-day eagerness means it will fill in gaps with its general knowledge, which might have an absence of perfect alignment with your specific hotel's detailed service recovery protocols. The more informal your prompt, the more the AI has to guess what a good manager (you!) would want.

Second, and far more powerful for most professional tasks, is structural prompting. This is like giving your new remote associate a detailed, well-thought-out project brief. You are providing a comprehensive set of instructions, context, examples, and desired outcomes. You are carefully outlining the what, why, and how. Crafting a structural prompt requires more initial time and thought from you, the manager. You need to be clear about your objectives before you instruct your AI. However, this upfront investment in clarity almost invariably yields results that are significantly more accurate, more strategically aligned, and more immediately usable, saving you substantial revision time later. This detailed briefing is how you empower your AI associate to perform at its best.

Universal Keys to Effective AI Management

Whether your interaction is a quick conversational check-in or a detailed structural briefing for your AI remote associate, three universal principles ("Golden Triangle") will dramatically improve the quality of its work: specificity, context, and examples.

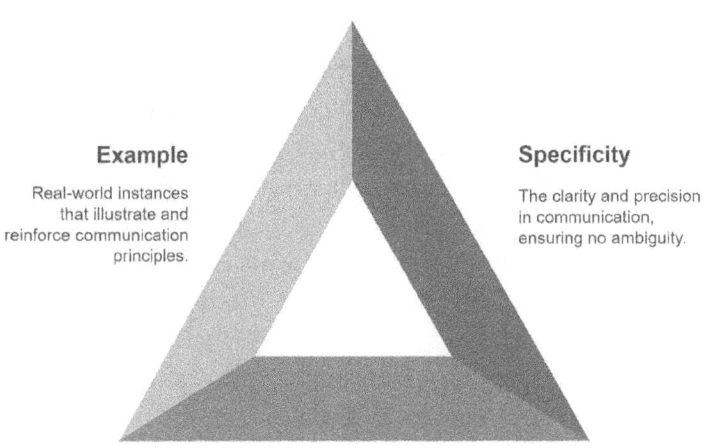

1. **Specificity**: Clearly Define the Task for Your "New Hire."

   Just as you would avoid telling a new human employee, "Just write some marketing stuff," you cannot expect great results from a vague AI prompt. The more precise you are in articulating the exact task, deliverables, parameters, and desired outcome, the better your AI associate can perform. It reduces guesswork and aligns the AI's efforts with your specific needs.

   - *Less Effective (Vague instructions for your AI associate)*: "Guest concierge, write something about our hotel's new spa services for the website."
   - *More Effective (Clear, specific briefing)*: "Guest concierge, please write one paragraph of about 150 words for the 'Wellness' section of The Serene Shores Resort website. This paragraph should introduce our new Ocean Harmony Ritual, a ninety-minute signature spa treatment. Your description must highlight: 1. The use of locally sourced, organic seaweed and marine botanicals. 2. The unique mix of gentle exfoliation, a warm mineral-rich wrap, and a rhythmic wave-inspired massage. 3. The main benefits for guests are deep relaxation, skin rejuvenation, and a strong sense of connection to the ocean. The tone must be luxurious, serene, and deeply restorative, aimed at guests looking for an exclusive, high-end wellness experience. Please make sure the language matches our brand voice guidelines (see attached SereneShores_BrandVoice.pdf)."

   This level of detail is like giving your AI associate a perfect first assignment brief, setting it up for success.

2. **Context**: Provide All Necessary Background Information.

   On its first day, your AI remote associate knows nothing about your hotel's current promotions, a specific guest's history, or an ongoing operational challenge unless you provide that information. AI, especially general-purpose models, lacks inherent knowledge of your unique, proprietary business context.

- *Less Effective (Lacking context for your AI associate)*: "Guest concierge, draft an email to a guest who was unhappy."
- *More Effective (Providing full situational context)*: "Guest concierge, we need to draft an empathetic and solution-oriented email to our Gold Loyalty member, Mrs. Davis. She stayed with us last weekend (Booking ID #G12345). She reported in her post-stay survey (see attached Davis_Survey_May15.pdf) that although she loved her room view, she was significantly inconvenienced by the elevator being out of service for three hours on Saturday morning. Our records show this was due to emergency maintenance. Our goal is to sincerely apologize, validate her frustration, and offer a gesture of goodwill to encourage her to return, perhaps a 15% discount on her next stay or 5,000 bonus loyalty points. Please draft the email from me, [Your Name], hotel manager, ensuring the tone is one of genuine apology and proactive care, reflecting our commitment to guest satisfaction. Refer to our standard Service Recovery Guidelines (document ServiceRecovery_Protocols_v2.docx) for appropriate language and offer parameters."

Giving your AI associate this rich context is like providing a thorough briefing, enabling it to respond intelligently and appropriately.

3. **Examples** (Few-Shot Prompting): Show Your AI Associate What Success Looks Like.

For new human employees, showing them examples of excellent past work is often the fastest way to get them up to speed on quality standards and stylistic expectations. The same is true for your AI. Instead of only describing your desired output, show one or more high-quality examples. This is known as few-shot prompting.

- *Less Effective (An absence of examples for your AI associate)*: "Guest concierge, generate some social media captions for our summer family package."
- *More Effective (Showing examples)*: "Guest concierge, please generate three new, engaging Instagram captions for The Seaside Inn's Summer Family Splash package. Our brand voice is fun, friendly, and focuses on family memories. Each caption should be around twenty to thirty words and include two or three relevant emojis and three to five family-travel hashtags. Here are two examples of past successful captions for similar family packages:

    Example 1: Sun, sand, and smiles! Our Family Splash package has all you need for a great beach trip. Dive into fun! #SeasideInnFamily #SummerFun #BeachVacay #FamilyMemories #KidsLoveIt'

    Example 2: Making memories one sandcastle at a time! Our beachfront suites and kid-friendly activities make The Seaside Inn your perfect family escape. Let the adventures start! #FamilyGetaway #BeachKids #HotelDeals #MakingMemories #TravelWithKids'

  Now, using that same voice and style, create three new captions specifically highlighting these features of the Summer Family Splash package: 1. Our newly renovated kids' pool with a waterslide. 2. The complimentary daily ice cream vouchers for children under twelve. 3. The early booking discount of 15% if booked by June 1."

By providing these gold standard examples, you are giving your AI associate a clear benchmark for quality and style, significantly improving the relevance and usability of its initial drafts and saving considerable time in revisions. Consistently applying specificity, context, and examples when managing your AI interactions will fundamentally elevate your collaboration, turning your AI from a generalist tool into a highly effective, specialized remote associate.

Core Prompting Frameworks

Although the "Golden Triangle" provides universal principles, structured frameworks can act as reliable briefing templates for your AI associate, especially for recurring communication tasks or when aiming for consistent output across your team. Mastering a few will significantly enhance your prompting effectiveness, ensuring your associate understands the full scope of their assignment.

- The APE Framework (action, purpose, expectation) is good for straightforward content creation, where you need to clearly define what the AI should do, why it is doing it, and what a good result looks like.

    - **Action**: State the verb that defines what you want the AI to perform, for example, write, summarize, list, translate, brainstorm, or compare.
    - **Purpose**: Explain the overarching why behind this request and the goal or strategic objective you are trying to achieve with this AI-generated output. Helping your AI associate understand the purpose leads to more relevant responses.
    - **Expectation**: Clearly outline your parameters for the outcome, perhaps an approximate length like

word count or number of paragraphs, the desired tone of voice, such as formal, warm, empathetic, or urgent, essential information that must be included, and the target audience.

    - *Hospitality Example (APE for a pre-arrival email for a Weekend Wellness Retreat package, briefing your AI guest communications associate)*: Action: "Write a personalized pre-arrival email." Purpose: "To build excitement for their wellness retreat, reconfirm package details, and offer appealing pre-bookable enhancements (like private yoga, a spa treatment, or a healthy gourmet cooking class) to make their experience even more special." Expectation: "The email should be 250–300 words. The tone must be warm and welcoming and convey serene, anticipatory calm. It must highlight core package inclusions (e.g., complimentary late checkout, healthy breakfast, spa credit). Enticingly list two to three optional enhancements with descriptions, pricing, and a clear booking link or contact information. Ensure the communication uses our luxury wellness brand voice, emphasizing tranquility and rejuvenation."

- The RACE Framework (role, action, context, expectation) is more comprehensive, particularly when you need your AI to adopt a specific persona, viewpoint, or level of professional authority, just like you would brief a specific team member.

    - **Role**: Who should the AI convincingly embody? Examples include your hotel's general manager, a knowledgeable concierge, a persuasive sales manager, an empathetic guest service agent, or the hotel brand itself.

- **Action**: With all its specific requirements, what precise communication task do you need the AI to perform in this defined role?
- **Context**: What critical background information, specific situational details, or essential historical data must the AI understand thoroughly to act effectively?
- **Expectation**: What is your clearly defined desired outcome? Articulate the ideal tone, essential content, specific formatting, length constraints, and any call to action.
- *Hospitality Example (RACE for a director of catering following up with a potential wedding client concerned about vegetarian/vegan menu options, instructing your AI event sales associate)*: Role: "You are the highly experienced, creative director of catering at The Heritage Grand Ballroom, a landmark luxury hotel venue renowned for flawless high-profile events and bespoke five-star dining." Action: "Draft a personalized, thoughtful, and reassuring follow-up email to our important prospective wedding client, Ms. Evelyn Hayes, regarding her site tour yesterday." Context: "Ms. Hayes loved the venue but expressed concern that our standard vegetarian/vegan options seemed limited for her 200 guests (30% vegetarian/vegan). Our executive chef, Jean-Pierre Dubois, is internationally acclaimed for innovative plant-based cuisine, a fact that may have an absence of full conveyance." Expectation: "The email must be profoundly reassuring, proactive in addressing her concern, and elegantly persuasive about our culinary excellence in this area. Warmly acknowledge her positive feedback on the venue. Address menu concerns with empathy and confidence. Emphasize chef Dubois's expertise and our commitment to

an extraordinary culinary experience for *all* guests. Enthusiastically invite Ms. Hayes for an exclusive, complimentary private menu tasting focused on innovative vegetarian/vegan dishes. The tone must be of sophisticated problem-solving and gracious hospitality. Conclude with my direct contact information so she can arrange this special tasting."

- The TAG Framework (task, action, goal) is excellent for generating clear, actionable operational content like standard operating procedures (SOPs), internal process guidelines, detailed checklists, or foundational training materials.

    - **Task**: What is the overall objective, specific operational process, or key area of hotel functioning that needs to be clearly defined or documented?
    - **Action**: What are the precise, sequential steps, approved procedures, essential content elements, or non-negotiable best practices required to complete this task or manage this process to hotel standards?
    - **Goal**: What is the ultimate intended outcome of performing this task or following this SOP? What is the desired standard of operational performance or guest service excellence?
    - *Hospitality Example (TAG for a new front desk SOP for handling guest lost-and-found inquiries, instructing your AI operations manual drafter)*: Task: "Develop a new, comprehensive, easy-to-follow SOP for effectively managing all guest lost-and-found inquiries at The Cityscape Grand (300-room upscale hotel)." Action: "Write a detailed, step-by-step procedure for our front office SOP manual. It must cover: 1. Professionally receiving the initial inquiry (in-person, phone, email). 2. Accurately logging all critical lost item details (description, estimated

location, date/time, guest contact information). 3. Secure storage and tagging protocols for found items. 4. The defined internal search process (contacting housekeeping, F&B, security CCTV review for high-value items if appropriate). 5. Guidelines for timely, professional guest notification if the item is located. 6. Approved secure procedures for guests to reclaim items or for arranging secure shipping at guest expense." Goal: "Ensure every lost item inquiry is handled with operational consistency, genuine empathy, and maximum efficiency, maximizing successful reunions, minimizing hotel liability, and reinforcing our hotel's reputation for exceptional care and proactive problem-solving."

- The IDEAS Framework (intent, data, example, audience, style) is powerful for crafting impactful marketing communications, sales collateral, or any communication where deep audience understanding and achieving a specific persuasive intent are key. This is like giving your AI marketing associate a full campaign brief.
    - **Intent**: What is the core strategic purpose of this communication, for example, to drive bookings, build brand awareness, announce news, or encourage loyalty?
    - **Data**: What key facts, figures, operational details, or specific information must be accurately included for the communication to be correct and achieve its intent?
    - **Example**: If you have a previous communication like a successful email campaign or engaging social post that achieved a similar intent or resonated with a similar audience, provide it as a model.

- **Audience**: Who are you trying to reach and influence? Be detailed. Include demographics, such as age or location; psychographics, like lifestyles, values, or interests; likely needs or pain points; travel motivations; and current stage in their guest journey with your brand.
- **Style**: What specific tone of voice, overall emotional feeling, and presentation format are most appropriate and effective for this intent and this audience, for example, playful, luxurious, urgent, or informative?
- *Hospitality Example (IDEAS for an email campaign for a new Winter Wellness Weekend spa package, briefing your AI marketing associate)*: Intent: "Generate immediate excitement and directly drive at least fifty early bird bookings for our new Winter Wellness Weekend package (offered Jan–Feb only)." Data: "Package includes two-night deluxe king stay, healthy breakfast, $150 spa credit per person, thermal suite access, guided winter nature hike. Early bird price: $799/couple (20% saving)." Example: "For style and persuasive feel, refer to our successful Summer Serenity Escape campaign (Subject: Your Personal Oasis Awaits), which achieved a 25% open rate and 5% click-to-booking conversion." Audience: "Discerning, affluent couples, thirty to fifty-five, interested in wellness, nature, luxury escapes. Often planning romantic getaways, anniversaries, and stress-relief breaks. Typically book four to six weeks out." Style: "Exceptionally inviting, cozy, evocative of winter warmth and rejuvenation. Emphasize renewal, luxurious escape, and tranquility. Use rich, sensory language. Visually convey warmth and serenity. Compelling call to action: 'Book Your Winter Rejuvenation Now and Save 20% Before This Offer Melts Away!' linking directly to the package booking page on our website."

- The START Framework (situation, task, action, result, tone) is invaluable for structuring compelling narratives, like training case studies, stakeholder success stories, or talking points for presentations. This helps your AI training content developer or AI communications specialist build a story.
    - **Situation**: Briefly set the scene: the initial context, challenge, or opportunity.
    - **Task**: The specific objective that needed to be accomplished or the critical issue to be resolved.
    - **Action**: The precise steps or initiatives taken to address the task or challenge.
    - **Result**: The tangible, measurable outcome or positive impact of those actions. Quantify these results whenever possible, for instance, "A 15% increase in guest satisfaction scores" or "A savings of eight staff hours per week."
    - **Tone**: The overall feeling, narrative style, or key underlying message the AI should adopt, such as inspiring, cautionary, analytical, or celebratory.
    - *Hospitality Example (START for an internal training case study on service recovery, instructing your AI training content developer)*: Situation: "Busy, fully booked Saturday night, peak season at The Metropolis Grand. Major water pipe burst on the third floor, impacting three occupied rooms, minor hallway flooding." Task: "Night manager Ms. Sharma needed to quickly, calmly relocate affected guests, manage their distress, coordinate emergency engineering and housekeeping, and minimize disruption to other guests." Action: "Ms. Sharma personally visited rooms and spoke empathetically with guests. She quickly found and blocked alternative upgraded rooms and offered sincere apologies and appropriate service recovery (F&B vouchers, points). She

constantly coordinated with engineering (to contain leaks/repairs) and housekeeping (to clean/room prep). She carefully documented actions and provided comprehensive handover." Result: "Despite the incident, two of the three affected guests praised Ms. Sharma's calm, professionalism, and empathy in online reviews. There were no broad negative online reviews about the incident. The damage was contained, and repairs were initiated within two hours with minimal disruption." Tone: "Write a concise, inspiring, instructional case study for leadership development and guest service training. Emphasize problem-solving under pressure, grace, professionalism, team coordination, and guest empathy."

Mastering these five frameworks will give you and your team a robust toolkit for briefing your AI associates effectively across a wide range of hospitality tasks, saving time and improving output quality.

Advanced Prompting Techniques

Once you are comfortable managing your AI associate with core frameworks, you are ready for advanced techniques to tackle more complex tasks and drive more detailed responses from your AI.

- **Chain-of-Thought (CoT) Prompting**: Guiding Your AI Associate to Think Step-by-Step and Clearly Show Its Work

  When you are faced with a complex problem that requires multiple logical steps or when it is vital to understand how your AI associate arrived at a conclusion, use chain-of-thought (CoT) transparency prompting. Instead of simply asking for the final answer, you strategically instruct it to think step-by-step and explicitly outline its reasoning process, essentially showing its work. This is like asking a new human associate for their methodology and rationale, not just the final report.

CoT prompting often leads to more accurate, reliable, and well-reasoned outputs because it forces the model to break down complex problems. It also provides transparency into how the AI is "thinking," allowing you to see its connections. Observing how AI dissects a complex problem can also be a helpful tool for your professional learning.

- *Hospitality Example (Elaborated strategic analysis with CoT for your AI revenue analyst associate, understanding a dip in Q4 weekend leisure bookings)*: A simple instruction would be: "Ronnie (your AI revenue analyst), why are our Q4 weekend leisure bookings suddenly so slow?" A far more powerful CoT instruction is: "Ronnie, your designated role is our exceptionally experienced director of revenue management for The Grand Metropolis. Your critical task today is to help me understand precisely why our weekend leisure bookings for the upcoming Q4 period (October–December) are pacing 15% behind forecast and last year. I need you to adopt a rigorous, step-by-step analytical approach. Please think aloud and clearly show me your work by addressing each of the following points sequentially: First, carefully review and summarize our current occupancy on the books for all Q4 weekend dates from Q4_PacingReport_May2025.xlsx. Compare this to last year and our three-year average. State key findings. Second, using web search, identify major new city-wide events, conferences, or competitor hotel openings/refurbishments for Q4 that might impact our leisure demand. List these and their potential impact. Third, carefully analyze our current Q4 weekend leisure package offerings and pricing strategies detailed in Q4_LeisurePackages_PricingStrategy.pdf. Compare these against the current public offers from our top three local competitors (The Plaza Hotel, The Regent Arms, and The Carlyle Grand; use web search). Summarize competitive pressures or opportunities. Fourth, consider relevant broader macro-environmental factors (regional airline capacity, consumer travel sentiment reports, local tourism initiatives) influencing Q4 leisure travel to our city. Explain their potential relevance. Finally, based on this comprehensive, step-by-step analysis, summarize your top three most plausible hypotheses for our Q4 weekend leisure underperformance. For each hypothesis, explain your reasoning, reference key data from steps one through four, and suggest one specific, actionable strategic recommendation for our commercial team to stimulate demand."

This detailed CoT brief skillfully guides the AI associate through complex analysis, ensuring thoroughness.

- **Hierarchical Prompting**: Guiding Your AI Associate Through Large, Complex Document Creation

  Attempting to generate an extensive document like an SOP manual in one single prompt can lead to muddled results or exceed the context window. Hierarchical prompting, breaking the project into phases, is much more manageable and effective.

  1. **Begin by Prompting for a Detailed Outline**: This is your project blueprint. Collaborate with your AI associate on an outline. For an SOP manual, list

major operational areas and ask the AI to suggest a logical flow.

2. **Develop Each Key Section Iteratively**: Once the outline is final, focus on developing content for each major section individually, treating each as a mini-project. For the SOP manual, you might instruct the AI to draft "Section II: The Complete Guest Check-In Process," providing sub-points, then review and refine it.
3. **Synthesize and Polish the Entire Document**: Once all sections are complete, prompt your AI to help write an introduction, generate a table of contents, and craft a concluding summary. This layered approach ensures coherence and quality for large projects.

Advanced Reasoning

While the prompting skills we've covered are excellent for directing AI to perform specific, execution-oriented tasks, a more advanced level of collaboration is needed when you face complex strategic challenges. Standard prompting is the perfect tool when you know the output you want. For example, you would use it to "draft a welcome email for a VIP guest" or "summarize the attached meeting notes into five bullet points." You are providing clear instructions for a defined task. However, for challenges that are ambiguous, multifaceted, and lack a single "right" answer, you need to engage the AI as a thinking partner who can help you analyze a situation and evaluate potential solutions.

This is the core of advanced reasoning. It marks the shift from telling the AI *what to write* to providing it with a framework for *how to think*. Consider this analogy to make this distinction tangible: Standard prompting is like giving a talented cook a detailed recipe. You specify every ingredient, measurement, and step, and they will execute it flawlessly to produce the dish you designed.

Advanced reasoning, on the other hand, is like giving an executive chef a strategic challenge: "We need a new signature dessert for our fall menu that uses local apples, appeals to our luxury clientele, and has a food cost below 20%." You are providing the goals, context, and constraints and empowering the chef to use their expertise to create the recipe itself.

Structured reasoning frameworks are required to facilitate this deeper, more analytical collaboration with your AI. These are more like a comprehensive consultant's brief or a strategic blueprint that you fill out. They provide the necessary guardrails for the AI's powerful capabilities, ensuring its "thinking" process is logical, thorough, and aligned with your business objectives.

Strategic Reasoning Frameworks for Complex Hospitality Decisions

For particularly complex strategic evaluations, dedicated reasoning frameworks can guide even advanced AI to produce comprehensive analyses. Two such powerful frameworks are CASINO and SCATE.

- The CASINO Framework (context, alternatives, standards, inference, notions, outcomes) is well-suited for multifaceted strategic challenges with multiple potential paths. Imagine your leadership is considering repositioning an aging airport hotel facing new competition.

  You could prompt a reasoning AI using CASINO: "Please analyze our 250-room airport hotel's current challenging market position and evaluate potential repositioning strategies using the CASINO method. **Context**: Our hotel, The Airport Gateway Inn, has historically served transient business travelers and airline crew, focusing on 'reliable convenience.' We now face new competition from The Vibe (lifestyle) and StayLong Suites (extended-stay). Our RevPAR index declined 8% last quarter. **Alternatives**: Evaluate repositioning as: 1. A wellness transit hub (spa, fitness, healthy

F&B). 2. A hyper-efficient tech hub (contactless journey, co-working). 3. A curated gateway to the region (leisure, local experiences). **Standards** (Evaluation Criteria): Evaluate each against projected RevPAR impact, brand alignment, estimated operational feasibility/investment, and potential for sustainable differentiation. **Inference**: According to local market demand (corporate, 'bleisure,' wellness) and competitor analysis, determine which alternative offers the strongest potential for sustainable advantage and improved finances. **Notions** (Challenging Assumptions): Explicitly challenge our assumption that airport hotels primarily target only short-stay business/crew. What if trends like extended leisure layovers or work-from-hotel represent untapped opportunities? **Outcomes**: For your recommended strategy, project its likely impact on the guest mix, target ADR, guest satisfaction, and new staff skills over three years."

Dissecting the CASINO Prompt for The Airport Gateway Inn: The context grounds the AI in the hotel's specific reality. Alternatives provide bounded options, focusing on the AI's comparative analysis. Standards define criteria for judging alternatives, ensuring evaluation is not arbitrary. Inference instructs the AI on its core analytical deduction, synthesizing information to infer which alternative is strategically superior. Notions push the AI beyond surface analysis, questioning preconceived ideas to surface innovative insights. Finally, outcomes requires the AI to look forward, outlining tangible results and implications, moving from analysis to actionable foresight. This detailed brief structures the AI's "thinking," ensuring comprehensive, well-reasoned recommendations.

- The SCATE Framework (situation, complications, analysis, tradeoffs, execution) is ideal for evaluating immediate business decisions with distinct options and significant pros and cons. You could prompt your AI collaborator using SCATE: "Using the SCATE method, please provide a comprehensive evaluation of whether our 400-room luxury beachfront resort, The Coral Sands, should introduce a new, mandatory $35 daily resort fee. **Situation**: We currently offer complimentary amenities (premium beach service, Wi-Fi, fitness classes), costs absorbed in room rates. Our positioning is exceptional experience/value, lacking the lowest price. **Complications**: OTAs display base rates first, potentially making us look more expensive if amenities are bundled versus unbundled. A general consumer pushback against hidden fees exists. Some local competitors have resort fees. **Analysis**: Analyze potential net revenue gain from the fee (factoring occupancy, length of stay) versus risk of booking volume loss. Analyze existing guest feedback (GuestFeedbackSummary.pdf attached) for amenity usage and perceived value. Research and summarize resort fee structures/inclusions for our top five local competitors. **Tradeoffs**: Evaluate key tradeoffs: revenue increase versus potential negative impact on guest satisfaction, booking conversion, online review sentiment, rate transparency, and guest trust. **Execution**: If recommending the fee, outline a guest-centric implementation plan: specific amenities covered, clear communication at all booking stages, staff training, and how financial impact/guest sentiment will be measured."

Dissecting the SCATE Prompt for The Coral Sands: This prompt guides the AI through a structured business case evaluation. The situation component establishes the baseline facts for the AI, defining the resort's current amenity structure, cost absorption model, and market positioning. Complications then introduce the key challenges and external pressures influencing the decision, such as how OTAs display pricing, general consumer sentiment about fees, and the actions of local competitors. The analysis section provides the AI with specific, direct instructions for data gathering and processing, asking it to calculate potential revenue impact versus booking risk, analyze guest feedback from a provided document, and research competitor practices. Following this, tradeoffs requires the AI to move from pure analysis to a balanced evaluation, explicitly weighing the potential financial gains against the risks to guest satisfaction, rate transparency, and brand trust. Finally, the execution component makes the AI's output highly actionable by instructing it to outline a guest-centric implementation plan if it recommends moving forward with the fee, covering communication, training, and measurement.

The above frameworks with advanced reasoning models help ensure the AI considers all angles and important data points. It provides valuable strategic guardrails for the AI's reasoning, guiding it toward more focused, relevant, and trustworthy output.

The choice between using the CASINO or SCATE framework depends on the scope and nature of the decision you are asking the AI to help you evaluate.

You should use CASINO when you are facing a large, multifaceted strategic challenge that involves exploring and comparing several potential long-term paths for your business. It is best for what-if scenarios that could fundamentally change your hotel's identity or direction.

Think of using CASINO for questions like:

- "What should the new identity and market position of our recently acquired, aging hotel be?"
- "What should our five-year growth strategy look like: Should we expand into new markets, acquire a competitor, or focus on maximizing value from our existing asset?"
- "How should we fundamentally reinvent our guest experience to appeal to a completely new demographic?"

CASINO is your tool for high-level, exploratory strategic planning where you need to weigh distinct, comprehensive future visions against each other.

Use the SCATE Framework for specific, tactical decisions. You should use SCATE when evaluating a more immediate and specific business decision with a clearer set of options and significant pros and cons. It is ideal for analyzing a single proposed action or making a go/no-go choice on a well-defined project.

Think of using SCATE for questions like:

- "Should we implement this new, mandatory $35 resort fee?"
- "Should we invest in renovating the lobby this year or upgrade the PMS instead?"
- "Should we outsource our laundry services to Vendor A versus keeping them in-house?"

SCATE is your tool for focused, operational, and financial evaluations where you need a comprehensive analysis of the potential impacts and tradeoffs of a specific, actionable choice.

In short, a simple way to decide is to ask yourself: "Are we deciding *what our future could be* (use CASINO), or are we deciding *if we should take this specific action now* (use SCATE)?"

Prompting as a Core Leadership Skill: The Ongoing Human-AI Conversation

In the AI-augmented hospitality field, your team's ability to communicate with clarity, context, strategic intent, and creative insight with your AI remote associate is becoming an important leadership competency and a basic operational skill. It involves learning to ask better questions, provide more complete briefings, critically evaluate complex information from a non-human source, and guide these systems toward results that match your hotel's values and goals.

The new remote associate mindset is central. Specific AI tools and prompting techniques will change quickly; that is AI's nature today. Yet, the basic principles of clear human communication, sharp strategic thinking, iterative refinement, and human-centered design in your AI interactions will last. True skill comes from consistent practice, thoughtful experimentation, continuous learning, and engaging in dynamic conversation with AI and your teams as you innovate together.

Creating a Shared Prompt Library

You have explored prompting frameworks and advanced techniques. But how do you ensure these skills translate into consistent, high-quality AI outputs across your team, saving everyone time? The answer lies in collaboratively building and maintaining a shared prompt library. Think of this as your hotel's curated collection of gold-standard AI assistant instructions covering common tasks. It is a practical next step, changing individual skills into collective capabilities. Instead of each team member reinventing the wheel, they can draw from pre-vetted, optimized prompts.

Investing time in a shared prompt library offers several benefits:

- It ensures brand voice and operational consistency.
- It accelerates AI adoption and proficiency as new or less confident team members can quickly become effective.
- It saves significant time by eliminating redundant effort.
- It captures and shares best practices, becoming a living repository.
- It facilitates continuous improvement as teams collectively refine prompts.

Building Your Departmental Prompt Library

Creating a valuable prompt library is an ongoing collaborative effort. Here is how departments can systematically collect, refine, and share effective prompts.

1. **Collect Promising Prompts**: Encourage team members to submit prompts that yield good results for common tasks. A front desk team might submit prompts for guest requests; sales might contribute proposal email prompts. Initially, target prompts for regular tasks where AI offers clear benefits.
2. **Refine and Standardize for Excellence**: Designate AI champions or a working group to review submissions or make it a recurring team meeting agenda item. Ensure prompts incorporate principles like specificity, context, and examples and leverage relevant frameworks (APE, RACE, etc.). For each prompt, include a title/purpose, targeted AI tool(s), key variables/placeholders, and desired outcome or example output. Verify prompts are designed to avoid biased or inappropriate content and include data privacy reminders, for instance, "Remember to remove any actual guest PII before pasting into a public AI tool unless using an approved enterprise platform." This aligns with AI governance.
3. **Share and Organize for Easy Access**: Choose an easily accessible platform like a shared drive (Google Docs, SharePoint), internal knowledge base, or a dedicated channel in Slack or Microsoft Teams. Structure the library intuitively, perhaps by task type, department, or AI tool. If possible, enable search functionality.
4. **Maintain and Develop Continuously**: AI models change, so schedule periodic reviews (monthly/quarterly) to update, add, or remove prompts. Encourage user feedback and suggestions. Recognize contributors to foster shared learning.

By systematically developing shared prompt libraries, hotel departments can boost collective AI proficiency, ensure consistent quality, and unlock substantial time savings, moving prompting from an individual skill to a powerful team asset.

Key Takeaways:

- Adopting the new remote associate mindset is essential for effective AI interaction, treating AI like a new team member requiring clear guidance, context, and examples (the "Golden Triangle" of specificity, context, and examples).
- Using structured prompting frameworks (like APE, RACE, TAG, IDEAS, and START) provides reliable briefing templates for common hospitality tasks, helping save time and achieve more consistent, high-quality AI outputs.
- Employing advanced prompting techniques such as chain-of-thought (CoT), hierarchical prompting, and strategic reasoning frameworks (CASINO and SCATE) elevates AI collaboration to sophisticated co-creation and strategic analysis.
- Establishing a shared prompt library is a team best practice that captures and disseminates effective prompts, ensuring brand consistency, accelerating AI adoption, saving time, and fostering continuous improvement.

Bringing This into Focus:

- **Practice the New Remote Associate Mindset This Week**: For your next significant interaction with an AI tool, consciously prepare and brief the AI as if it were a new, capable team member. Provide clear context, specific instructions, and examples. Notice if this improves the AI's output quality.
- **Apply One Core Prompting Framework**: Choose one core prompting framework from this chapter (APE, RACE,

TAG, IDEAS, or START) relevant to a task you need to accomplish. Use it to structure a detailed prompt for AI assistance. Evaluate how the framework helped clarify your request.

- **Experiment with an Advanced Prompting Technique**: For a complex analytical question, strategic decision, or longer document, try using chain-of-thought, hierarchical prompting, or a strategic reasoning framework (CASINO or SCATE). Observe how the technique helps manage task complexity.

Leadership Action Agenda:

- **Host a Prompting Frameworks Workshop**: Dedicate a team meeting to introduce one or two core prompting frameworks and one strategic framework. Have team members practice applying them to a common departmental task or strategic scenario, then share prompts and outputs for peer feedback. This investment builds practical skills.
- **Initiate Strategic Reasoning with AI Simulation**: Select a current strategic challenge with your leadership team. Use the CASINO or SCATE reasoning framework to collaboratively structure a detailed prompt for an advanced AI model. Analyze the AI's output as a team, discussing its insights and limitations. Reflect on how this AI-assisted process differed from your usual approach.
- **Champion Effective Prompting and a Shared Prompt Library**: Actively encourage and recognize team members who craft effective prompts. Task AI champions or a departmental working group with initiating a shared prompt library for common tasks, establishing a process for collecting, vetting, and sharing successful prompts to foster peer learning and operational consistency.

6
USING AI WITH OFFICE 365/GOOGLE SUITE

ALTHOUGH MANY OFFICE applications are starting to integrate AI directly, you don't need to wait to leverage its power. By learning how to effectively prompt general AI models, you have 24/7/365 access to a powerful assistant that can help you generate content, ideas, and frameworks for your documents, spreadsheets, presentations, and more.

This chapter provides numerous sample prompts to illustrate this process. Viewing these as foundational starting points rather than rigid scripts is essential. To achieve the best and most relevant results for your specific needs, you should customize and expand upon these examples by applying the comprehensive prompting frameworks we discussed in the previous chapter.

The key insight arrives when you recognize that you can gain a substantial head start on complex tasks by having a well-structured conversation with an AI. You then take the AI-generated output, whether it is an initial draft, a template outline, an analytical framework, or a meeting synopsis, and transfer it into your familiar office applications for your expert refinement, customization, and implementation.

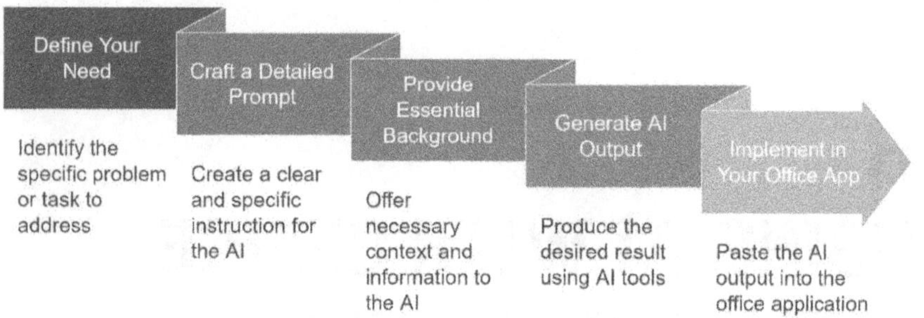

The core process involves several steps:

1. **Define Your Need**: What hospitality-specific task do you want to accomplish or improve?
2. **Craft a Detailed Prompt**: Ask your chosen AI model for the specific output you need.
3. **Provide Essential Background**: For the AI to generate relevant and useful output, you will often need to provide context. This might involve pasting text from existing documents, summarizing key data, providing a meeting transcript, or outlining specific parameters; examples are provided below each prompt.
4. **Generate AI Output**: The AI will provide a response based on your prompt and the background information.
5. **Implement in Your Office App**: Copy, paste, and adapt this output into your chosen software, such as word processors, spreadsheet programs, presentation tools, or email clients, or for your records from transcript analysis to fit your exact requirements.

Let's explore how you can prompt your way to efficiency and innovation across your office applications.

Word Processing Software (e.g., Microsoft Word, Google Docs)

Below are several ways you can use prompts with a general AI model with your word processing application. You can generate sophisticated content and frameworks by "talking" to the AI about your document needs, whether for creating new documents or refining existing ones. You then manually transfer this AI-generated output into your word processing software to finalize, customize, and utilize for your hospitality operations.

1. **Developing Enhanced Guest Profile Templates:**
   Engage a general AI model by describing your current guest profile system or pasting its structural elements. Ask the AI to help conceptualize a more sophisticated guest profile template, including suggestions for new personalization-focused sections or drafting example text for these. The AI's output can then be used to create a new or update an existing guest profile template document in your word processing software.

   - *Prompt Example*: "I need to create a new VIP guest profile template in Microsoft Word for our luxury property. I've pasted below the main sections of our current basic template and a list of data points we typically collect (e.g., stay history, F&B preferences). Based on this, help me outline an enhanced template. What new sections should I include for 'Anticipated Needs' or 'Personalized Pre-Arrival Touchpoints'? Please also draft some example text for a 'Standard Welcome Note' section, using our brand voice (formal yet warm), which we can then customize in the Word document based on specific guest details."

- *Examples of background information to paste/describe to AI:* Your hotel's brand voice guidelines, a description or pasted text of your current guest profile structure/data points, an overview of your property type and unique selling propositions, and goals for the new/updated Word document template.

2. Generating Stakeholder-Specific Versions of Policy Documents:

Attach the full text of an existing policy document to a general AI model into the chat interface. Instruct the AI to generate distinct variations of this policy that are personalized for different stakeholders. You can then copy each AI-generated version into separate word-processing documents for final formatting, review, and distribution.

- *Prompt Example:* "I've attached our complete updated employee data privacy policy. Please help me by drafting three distinct versions suitable for new Word documents: 1. A concise, easy-to-understand summary for all new hires (around 200 words). 2. A detailed version for departmental managers, highlighting their specific responsibilities. 3. A brief executive overview for senior leadership, focusing on key compliance points and potential risks."
- *Examples of background information to attach/describe to AI:* The full text of the current policy document, brief descriptions of each target stakeholder group and their information needs for each version of the Word document, and key objectives for each distinct document.

3. **Adapting Document Content for Multilingual Audiences:**

   Attach an existing English-language marketing brochure or guest service document into an AI model. Request assistance in adapting this content for international markets, ensuring it preserves brand voice while accounting for cultural distinctions. The AI-generated text can then be copied into your word processing software to create localized document versions.

   - *Prompt Example:* "Attached is the English text from our current in-room dining menu document. I need to create a new version of this menu in Spanish for our guests from Latin America. Please adapt the dish descriptions to be appealing and culturally relevant for this audience while maintaining our hotel's casual-upscale brand voice. The final output will be used to format a new menu document in Word."
   - *Examples of background information to attach/describe to AI:* The source language text from your document, detailed brand guidelines, information about the target markets for the new document versions, including key cultural considerations related to dining, service, or communication.

4. **Drafting SOPs with Decision Trees:**

   Share an existing standard operating procedure (SOP) by attaching its content into an AI model. Ask the AI to help you identify sections that could benefit from decision-tree logic for handling complex situations. The AI can help you draft the text and structure for these decision trees, which you can then integrate into your SOP document using your word processor's formatting tools.

   - *Prompt Example:* "I've attached the text of our current front desk SOP for handling guest room change requests. It's a bit vague. Can you help me draft the logic for a decision tree to include in this Word document? It should cover scenarios like the reason for the request (e.g., maintenance issue, preference), guest status (e.g., VIP, regular), and room availability. For each path, suggest appropriate actions or responses for the agent."

- *Examples of background information to attach/describe to AI:* The text of your current SOP document, a list of common complex scenarios related to the SOP, hotel policies on guest compensation or service recovery relevant to the SOP document.

5. **Outlining Competitive Intelligence Reports:**

Describe your key competitors and analytical objectives to a general AI model. Ask it to help you outline a comprehensive structure for a competitive intelligence report. This AI-generated outline can then serve as the framework for a new report you will write and format in your word processing software.

- *Prompt Example:* "I am tasked with writing a new quarterly competitive analysis report in Word for our city-center hotel. Our top three competitors are (Competitors X, Y, and Z). Please help me brainstorm a detailed outline for this report. What key sections should I include (e.g., pricing analysis, online review sentiment, new amenities/services, marketing campaigns)? For each section, suggest two or three specific data points or questions to research and include in the Word document."
- *Examples of background information to paste/describe to AI:* A list of primary competitors, key areas of competitive concern you want the report document to address, and examples of data sources you typically

use (e.g., STR reports, OTA listings, competitor websites).

6. **Analyzing and Revising Documents for Brand Voice Consistency:**

   Provide your hotel's brand voice guidelines to an AI model by attaching them. Then, paste text from a draft document (e.g., a marketing email or website copy) and ask the AI to analyze it for brand voice alignment. The AI can provide specific suggestions for rephrasing or editing, which you can then manually apply to your original word-processing document.

   - *Prompt Example*: "I've attached our hotel's brand voice guidelines, which emphasize a warm, inviting, and effortlessly elegant tone. Below that, I've pasted a draft of a new welcome letter Word document for our guests. Please review the draft letter and identify any phrases or sentences that don't quite match our brand voice. For each, suggest one or two alternative phrasings that would be a better fit."
   - *Examples of background information to attach/describe to AI*: Your detailed brand voice and style guide (attached text), the draft text from the word processing document you want to be reviewed, and information about the target audience for that specific document.

7. **Brainstorming Content for Guest Journey Documentation:**

   Describe a particular stage of your guest journey to an AI model. Ask for its assistance in brainstorming content ideas, key information to include, or personalization opportunities for documents related to that stage. Use the AI's ideas to then create or enhance guest journey maps or communication templates in your word processing software.

- *Prompt Example:* "We're creating a new document in Word to map our Post-Stay Engagement Journey. Our goal is to encourage repeat bookings and gather feedback. Can you help me brainstorm three or four key communication touchpoints for this post-stay document (e.g., thank you email, feedback survey, special offer for future stay)? For each touchpoint, suggest the primary message and a call to action we could then draft out in our Word document."
- *Examples of background information to paste/describe to AI:* A description of the specific guest journey stage you are focusing on for your document, your hotel's goals for that stage (e.g., increase loyalty, gather reviews), examples of current communications used.

8. **Drafting Initial Checklists for Regulatory Compliance:**

   Provide an AI model with the name or pasted text of a relevant regulation (e.g., local health and safety codes). Ask the AI to help you generate a preliminary checklist of key compliance actions or points. This AI-generated list can serve as a starting point for a more comprehensive compliance checklist you will finalize in a word processing document.

   - *Prompt Example*: "I need to create a new safety audit checklist document in Word for our hotel's pool area, based on the County Aquatic Facility Regulations, Section 3B (text pasted below). Can you help me extract five to seven key actionable items from this regulation that should be on our daily opening checklist? The final output should be a list I can easily copy into Word."
   - *Examples of background information to paste/describe to AI:* The specific text of the regulation or relevant sections, the operational area the checklist document

pertains to, and the frequency at which the checklist will be used (e.g., daily, weekly).

9. **Outlining and Structuring Content for Training Manuals:**

   Discuss a specific training need with an AI model. Ask for its help in brainstorming and structuring the content for a new training manual, including potential learning objectives, chapter or section headings, and practical exercise ideas. This AI-generated structure can then be used to develop the full training manual in your word processing software.

   - *Prompt Example*: "I'm tasked with developing a new training manual in Word for our new cohort of F&B servers. The topic is advanced upselling techniques for wine and appetizers. Can you help me outline the key sections for this manual? Suggest three main learning objectives. Also, brainstorm two or three common guest objections and brief ideal responses for a section on overcoming guest hesitations that we can then elaborate on in the Word document."
   - *Examples of background information to paste/describe to AI:* The target employee group for the training manual, specific skills or knowledge areas the manual needs to cover, and examples of successful upselling techniques currently used or common challenges servers face.

10. **Summarizing Long Reports into Briefs:**

   Attach a lengthy internal report or a collection of meeting notes into an AI model. Instruct the AI to generate a concise executive summary or a list of key action items. This summary can then be copied into a new word-processing document to create a quick brief for stakeholders.

- *Prompt Example:* "I've attached the full text of our twenty-five-page Annual Guest Satisfaction Trend Report below. Please generate a one-page executive summary (approximately 200–250 words) highlighting the top three positive trends, the top three areas for improvement, and key recommendations. I will then use this summary to create a brief in Word for our leadership team."
- *Examples of background information to attach/describe to AI:* The full text of the lengthy document, the intended audience for the summary document, the desired length or specific focus areas for the summary.

Spreadsheet Software (e.g., Microsoft Excel, Google Sheets)

Here are practical ways a general AI model can help you generate specific formulas, set up data manipulations, or design formatting rules, which you then build and execute in your spreadsheet software to solve common hotel challenges.

1. **Tiered Pricing Logic and Formula Generation:**

   Use an AI model to help you write complex nested IF statements or IFS formulas for dynamic room pricing based on multiple criteria, which you can implement directly in your pricing sheets.

   - *Prompt Example*: "I need an Excel formula for cell D2 to calculate a dynamic room price. The base rate is in cell A2. The occupancy percentage is in B2, and a special event flag (TRUE/FALSE) is in C2. The logic is: If occupancy > 90% OR special event is TRUE, increase the base rate by 25%. If occupancy is between 75% and 90%, increase the base rate by

15%. Otherwise, use the base rate. Help me write this formula."

- *Examples of background information to attach/provide*: A sample of your spreadsheet layout with cell references for base rate, occupancy, and event flags and clear rules for each pricing tier.

2. **Occupancy Forecasting with Weighted Averages and Confidence Intervals:**

   Ask an AI model to help generate formulas for forecasting occupancy using weighted averages of historical data and to calculate basic confidence intervals around these forecasts.

   - *Prompt Example*: "I have daily occupancy for the last three years in columns B, C, and D, starting from row 2. I want to forecast next year's occupancy in column E, weighting last year by 60%, two years ago by 30%, and three years ago by 10%. What Excel formula can I use in E2 and drag down? Also, suggest a simple formula to calculate a +/- 5% forecast variance range in columns F (Lower) and G (Upper)."
   - *Examples of background information to attach/provide*: A small sample of your historical daily occupancy data layout and the specific weighting percentages you want to use.

3. **Housekeeping Labor Calculation and Costing Formulas:**

   Use an AI model to develop formulas for calculating daily housekeeping staff needs based on room types and cleaning times and to look up pay rates for cost estimation.

   - *Prompt Example*: "Help me create Excel formulas. Column A: Room Type (e.g., Standard, Suite) B: Number of Rooms C: Minutes to Clean (This should be a VLOOKUP or XLOOKUP from a separate sheet named 'Standards' where Column A is Room Type and Column B is Minutes.) D: Total Minutes (B*C) E: Staff Needed (Total Minutes / 480, rounded up to whole number) F: Daily Cost (Staff Needed * 8 hours * VLOOKUP for pay rate from 'PayRates' sheet based on a role like Housekeeper). Provide the key formulas."

- *Examples of background information to attach/provide*: Examples of room types, the structure of your "Standards" and "PayRates" lookup tables, and standard shift length.

4. **Budget Variance Analysis Formulas and Conditional Formatting:**

   Have an AI model help generate formulas for budget variance calculations (amount and percentage) and suggest conditional formatting rules to highlight significant deviations.

   - *Prompt Example*: "In my Excel budget, column B is the budgeted amount, and column C is the actual amount. For row 5, what formula do I put in D5 for the variance amount (actual - budget)? And in E5 for variance percentage ((actual - budget) / budget)? Also, how can I set up conditional formatting in column E to show positive variances in green, negative variances in red, and highlight any variance greater than +/- 10% with yellow fill?"
   - *Examples of background information to attach/provide*: Your basic budget layout with relevant column headers and the specific thresholds for highlighting variances.

5. **Guest Feedback Keyword Counter and Sentiment Scoring Logic:**

   Use an AI model to suggest formulas for counting specific keywords in guest feedback text and for categorizing numerical sentiment scores with conditional formatting.

   - *Prompt Example*: "I have guest comments in Excel column A. How can I write a formula in column B to count (case-insensitive) how many times the word 'comfortable' appears in each comment? Separately, if column C has a satisfaction score from 1 to 5, what formula can I use in column D to categorize it as 'Poor' (1–2), 'Average' (3), or 'Good' (4–5)? Suggest conditional formatting for column D based on these categories."
   - *Examples of background information to attach/provide*: A few examples of guest comments and the scoring scale used for sentiment.

6. **Staff Schedule Overtime Calculation and Coverage Alerts:**

   Ask an AI model to help create formulas for calculating total weekly hours and overtime based on daily shift inputs and suggest conditional formatting for coverage issues.

   - *Prompt Example*: "My Excel staff schedule for row 2 has employee hours for Mon-Sun in cells B2 to H2. What formula in I2 sums these weekly hours? In J2, how do I calculate overtime hours if overtime is anything over 40 hours per week? Also, how can I apply conditional formatting to cell I2 to turn red if total hours are below a minimum of 30 for a full-time employee?"
   - *Examples of background information to attach/provide*: Your weekly schedule layout, rules for overtime

calculation (e.g., weekly threshold, daily threshold), and minimum staffing/hours rules.

7. **Inventory Reorder Point Formulas and Status Indicators:**

   Use an AI model to design formulas for calculating inventory reorder points and to create status indicators with conditional formatting for items needing attention.

   - *Prompt Example:* "For my F&B inventory in Excel: Column A = Item, B = Current Stock, C = Avg Daily Sales, D = Supplier Lead Time (days), E = Safety Stock Level. In column F, what formula calculates the Reorder Point as (Avg Daily Sales * Lead Time) + Safety Stock? In G, I need a formula that displays 'REORDER NOW' if Current Stock is at or below the Reorder Point, otherwise 'OK.' How do I conditionally format column G to highlight 'REORDER NOW' in red?"
   - *Examples of background information to attach/provide*: A sample of your inventory sheet layout and any specific logic for safety stock calculation if it's not a fixed number.

8. **Market Segment Profitability Formulas using SUMIFS/ PIVOT TABLES:**

   Have an AI model help develop formulas, such as SUMIFS, for aggregating revenue and costs by market segment to analyze profitability or suggest how to structure a Pivot Table for this analysis.

   - *Prompt Example*: "I have sales data: Column A = Booking ID, B = Market Segment (e.g., Corporate, Leisure, OTA), C = Revenue, D = Associated Variable Costs. I want to create a summary table that shows Total Revenue, Total Variable Costs, and Gross Profit (Rev-VC) for each unique Market Segment. Can you provide SUMIFS formulas to achieve this, assuming my unique segments are listed in cells F2, F3, F4, etc.? Alternatively, explain the steps to create a Pivot Table for this."

- *Examples of background information to attach/provide*: A sample of your sales data structure and a list of your defined market segments.

9. **Utility Consumption KPIs and Variance Formulas:**

   Use an AI model to conceptualize formulas for calculating key utility consumption KPIs, like KWh per occupied room, and for tracking variance against targets or historical data with conditional formatting.

   - Prompt Example: "In Excel, I have monthly KWh consumption in B2:B13 and occupied room nights for each corresponding month in C2:C13. What formula can I use in D2 (and drag down) to calculate KWh per occupied room night? If my target is 18 KWh/ORN, how can I use conditional formatting in column D to highlight months where consumption is 10% or more above this target?"
   - *Examples of background information to attach/provide*: Sample data layout for utility consumption and occupancy and your specific KPI targets.

10. **Data Structuring for Advanced Charts and Dashboards:**

   Ask an AI model for advice on how to structure data and what formulas to use (e.g., calculating percentage change, running totals) to prepare data for specific advanced chart types suitable for executive dashboards, including sparklines.

   - *Prompt Example*: "I have monthly guest satisfaction scores (1–100 scale) for the past twelve months in

Excel, cells B2:B13, with months in A2:A13. To prepare for a line chart showing trends, what formula can I use in C3 (and drag down) to calculate the three-month moving average? Also, how can I insert a sparkline in cell D2 that visualizes the trend of B2:B13? Suggest one other chart type suitable for showing this monthly score trend to executives."

- *Examples of background information to attach/provide*: A sample of the raw data you want to visualize and the key message you want the dashboard/chart to convey.

Presentation Software (e.g., Microsoft PowerPoint, Google Slides): Ten Advanced Applications

Leverage a general AI model to help you brainstorm narratives, structure content, and generate ideas for impactful presentations, which you then design and deliver using your presentation software.

1. **Strategic Narrative Presentations with Integrated Data Storytelling:**

   Use an AI model to develop compelling strategic narratives that integrate with data visualization, crafting a complete business story for your presentation software.

   - *Prompt Example*: "Help me develop a strategic narrative framework for our hotel's quarterly performance presentation to ownership. Weave financial results, market dynamics, and operational achievements into a cohesive story. Include guidance on integrating data visualizations that support the narrative and highlight key insights/future strategies."
   - *Examples of background information to attach/provide*: Key financial reports, a summary of recent

operational achievements/challenges, market intelligence reports, and the primary message/objective for the presentation.

2. **Competitive Set Analysis Presentations with Differentiation Mapping:**

   Ask an AI model to help create frameworks for competitive analysis presentations that visualize market positioning and strategic opportunities. These frameworks guide your presentation design.

   - *Prompt Example*: "Design a competitive analysis presentation framework for our boutique hotel. Visually map our position against three key competitors across price, service offerings, guest satisfaction, facility quality, and market perception. Include ideas for perceptual mapping and identifying opportunities/threats."
   - *Examples of background information to attach/provide*: Names of key competitors, available data on competitor pricing, services, reviews, and your hotel's unique selling propositions.

3. **Investment Proposal Templates with Financial Modeling:**

   Use an AI model to develop comprehensive capital investment presentation templates incorporating ROI analysis, risk assessment, and implementation planning. Build this structure in your presentation software.

   - *Prompt Example*: "Create an investment proposal presentation template for hotel renovation projects. Incorporate investment thesis, a summary of projected financial returns (ROI, payback), risk assessment, implementation timeline, and projected competitive advantage. Include sections for financial and operational stakeholders."

- *Examples of background information to attach/provide*: Details of the proposed investment project, outputs from financial models, and market research supporting the investment.

4. **Multi-Scenario Strategic Planning Presentations:**

   Have an AI model assist in building presentation frameworks that effectively communicate strategic planning across multiple potential scenarios. Develop these in your presentation software.

   - *Prompt Example*: "Develop a multi-scenario strategic planning presentation template for our resort (optimistic, expected, challenging conditions). Include scenario trigger indicators, adaptive response strategies for each path, and resource allocation frameworks."
   - *Examples of background information to attach/provide*: Your current strategic plan or forecast, definitions of the different scenarios and their potential impacts, and key strategic levers that can be adjusted.

5. **Guest Experience Journey Mapping Presentations:**

   Use an AI model to conceptualize visual presentations mapping the entire guest journey, highlighting touchpoints, emotional states, and enhancement opportunities. Visualize these in your presentation software.

   - *Prompt Example*: "Design a comprehensive guest journey mapping presentation for our all-inclusive family resort. Visualize the complete experience (discovery to post-stay). Include tracking of potential guest emotional states, pain point analysis, and 'moment of truth' differentiation opportunities."

- *Examples of background information to attach/provide*: Existing guest journey maps or process flows, guest feedback, survey results, online reviews, and descriptions of all services, amenities, and activities.

6. **Brand Development Strategy Presentations:**

    Ask an AI model to develop presentation frameworks for communicating brand development strategies, positioning shifts, and implementation roadmaps. Use the AI-generated structure in your presentation software.

    - *Prompt Example:* "Create a brand development strategy presentation template for our independent hotel shifting from traditional comfort to experiential, locally connected luxury. Communicate rationale, competitive positioning, target guest alignment, operational/service changes, marketing adjustments, and implementation timeline."
    - *Examples of background information to attach/provide*: Current and proposed new brand guidelines, market research on target guest preferences, and competitive analysis of experiential brands.

7. **Revenue Strategy Workshops with Interactive Components:**

    Use an AI model to design presentation frameworks for interactive revenue strategy sessions, combining data analysis with strategic decision points. Implement these in your presentation software.

    - *Prompt Example*: "Design an interactive revenue strategy workshop presentation. Combine performance analysis, market opportunity identification, and guided decision-making exercises. Include sections for segment strategy, channel optimization, and pricing positioning, with built-in discussion prompts."

- *Examples of background information to attach/provide*: Recent revenue performance reports, market forecasts, event calendars, and current pricing strategies.

8. **Training Program Design with Learning Objectives and Assessment Frameworks:**

   Have an AI model help create training program presentation structures aligning learning objectives, content delivery, and competency assessment. This guides your presentation software development.

   - *Prompt Example*: "Develop a structured training program presentation template for onboarding new front desk agents. Align learning objectives with content modules, delivery methodologies, and assessment frameworks. Include adult learning principles and knowledge retention strategies."
   - *Examples of background information to attach/provide*: Job description for front desk agents, list of essential skills/knowledge, and current SOPs for front desk operations.

9. **Operational Excellence Visualization with Process Mapping:**

   Use an AI model to brainstorm presentations that visualize operational processes, efficiency initiatives, and continuous improvement. Develop these ideas into slides in your presentation software.

   - *Prompt Example*: "Create an operational excellence presentation framework for our F&B department.

Visualize current processes, identify efficiency opportunities, and map improvement initiatives. Include ideas for process flow visualizations, bottleneck identification, and an implementation roadmap with KPIs."

- *Examples of background information to attach/provide*: Current process maps for F&B operations, data on operational performance, and identified pain points or inefficiencies.

10. **Crisis Management Planning and Communication Templates:**

    Ask an AI model to design presentation frameworks for crisis planning, response protocols, and stakeholder communication strategies. Build these out in your presentation software.

    - *Prompt Example*: "Design a comprehensive crisis management presentation template covering risk assessment for luxury hotels, response protocols by crisis type, communication frameworks, and recovery planning. Include stakeholder mapping, responsibility matrices, and scenario-based decision trees."
    - *Examples of background information to attach/provide*: Your hotel's current crisis management plan, a list of potential risks specific to your location/property, an organizational chart, and crisis team contacts.

Email (e.g., CRM, Microsoft Outlook, Gmail)

Use a general AI model to draft email sequences, communication plans, and response strategies, which you then personalize and send using your email client or CRM.

1. **Guest Communication Sequence Development:**

   Use an AI model to create sophisticated email sequences for the entire guest journey, then set up and personalize them in your email client or marketing platform.

   - *Prompt Example*: "Develop a comprehensive guest communication email sequence (pre-booking inquiry to post-stay feedback) for our luxury beach resort. Include messaging for inquiry follow-up, booking confirmation, pre-arrival anticipation, in-stay welcome/enhancements, and post-stay thank you/relationship development. Provide personalization variables for different guest types and booking channels."
   - *Examples of background information to attach/provide*: Your resort's brand voice and style guide, details of services, amenities, activities, different guest segments and their typical interests, and your current booking process.

2. **Sales Nurturing Sequence Frameworks:**

   Ask an AI model to build advanced prospect nurturing email sequences with adaptive messaging according to interaction patterns and lead qualification. Implement these via your email client or CRM.

   - *Prompt Example*: "Create a sophisticated sales nurturing email sequence for corporate meeting planners. Adapt messaging based on interaction engagement, planning timeline, and demonstrated interests. Include six sequence variations with decision tree logic for follow-up and objection handling."
   - *Examples of background information to attach/provide*: Profile of your ideal corporate meeting planner prospect. Your hotel's MICE offerings, unique

selling points, and common objections from meeting planners.

3. **Crisis Communication Templates with Stakeholder Segmentation:**

   Use an AI model to develop crisis communication frameworks with templated email messaging for different stakeholders and scenarios. Store and adapt these in your email client.

   - *Prompt Example*: "Develop a crisis communication email framework with templated messaging for various scenarios (power outage, security incident, health emergency, negative PR). Provide variations for affected guests, in-house guests, arriving guests, staff, ownership, and media. Include placeholders, escalation protocols, and timing guidelines."
   - *Examples of background information to attach/provide* include your hotel's overall crisis management plan, a list of potential crisis scenarios, key stakeholders, and communication preferences.

4. **Strategic Account Management Communication Plans:**

   Have an AI model help create sophisticated email communication strategies for key accounts, with frameworks for regular touchpoints and relationship development executed through your email client.

   - *Prompt Example*: "Design a strategic account management email communication plan for our top twenty corporate accounts. Incorporate cadence for touchpoints, frameworks for business review emails, methods for identifying/proposing growth opportunities, and strategies for reinforcing competitive positioning. Include templates."

- *Examples of background information to attach/provide*: Profiles of your top corporate accounts, your hotel's value proposition for corporate clients, and goals for strategic account management.

5. **Internal Communication Cascading Systems:**

   Use an AI model to develop frameworks for effective information cascading via email. An AI helps adapt messaging for different team segments, disseminated using your email client.

   - *Prompt Example*: "Create an internal email communication cascading system for major operational changes (new PMS, renovation). Ensure consistent core messaging while adapting delivery, detail, and guidance for executives, HODs, supervisors, and line-level staff. Include feedback collection ideas."
   - *Examples of background information to attach/provide*: Details of the operational change, organizational chart, and specific information needs of each audience level.

6. **Vendor Relationship Management Communication Frameworks:**

   Ask an AI model to help build systematic approaches to vendor communications via email, including performance reviews and issue resolution. These frameworks guide emails sent via your email client.

   - *Prompt Example*: "Develop a vendor relationship management email communication framework. Systematize performance reviews, issue resolution processes, and strategic development discussions. Include templates for quarterly reviews, service failure notifications, and partnership proposals."

- *Examples of background information to attach/provide*: List of key vendors and services, SLAs or contracts with vendors, and KPIs for vendor performance.

7. **Reputation Management Response Strategies:**

   Use an AI model to create frameworks for responding to guest reviews, which are often notified via email. These AI-generated responses are then personalized.

   - *Prompt Example*: "Design a comprehensive reputation management email response system for different review types (highly positive, positive with issues, mixed, negative, unfairly negative). Include email response templates that maintain our luxury brand voice, address failures, acknowledge praise, and outline recovery steps."
   - *Examples of background information to attach/provide*: Your hotel's brand voice and customer service philosophy, examples of different types of guest reviews, and staff empowerment guidelines for service recovery.

8. **Personalized Upselling and Cross-Selling Communication Frameworks:**

   Have an AI model help develop advanced approaches for revenue enhancement communications via email, blending personalization with strategic upselling. Send these via your email client or CRM.

   - *Prompt Example*: "Create a strategic upselling email communication framework for guests with existing bookings. Integrate personalization (segment, history, preferences), timing optimization, and psychological principles. Include methodologies and email templates for room upgrades, experience additions, and loyalty conversion."

- *Examples of background information to attach/provide*: Details of items available for upselling/cross-selling, guest segmentation criteria and data for personalization, and pricing for upsell items.

9. Media and PR Communication Strategies:

Use an AI model to build comprehensive frameworks for media relations communications via email. These strategies inform press releases and communications sent via your email client.

- *Prompt Example*: "Develop a media relations email communication playbook for our new boutique hotel. Include frameworks for proactive story pitching, reactive request handling, crisis management, and thought leadership. Include email templates for press releases, media alerts, pitch emails, interview briefings, and crisis statements."
- *Examples of background information to attach/provide*: Key messages and unique selling propositions, target media list, and profiles of key hotel spokespeople.

10. Sophisticated Email Management Workflows with Delegation Frameworks:

Ask an AI model to help design systems for managing high email volumes, including prioritization and delegation protocols. These systems guide actions within your email client.

- *Prompt Example*: "Design an advanced email management system for hotel leadership's high-volume inbox. Incorporate AI-assisted triage ideas, priority determination frameworks, response time standards, delegation protocols with accountability, and template utilization strategies. Suggest email client rules/features to support this."

- *Examples of background information to attach/provide*: Typical categories of emails received, current pain points in email management, and organizational structure for delegation.

Leveraging AI with Meeting Transcripts (From Tools like Zoom, Microsoft Teams, Google Meet)

Collaboration tools that record meetings and provide transcripts are invaluable. A general AI model can help you rapidly extract key insights, action items, and sentiment from these transcripts, saving significant review time.

1. **Comprehensive Meeting Summary Generation:**
   Use an AI model to condense a lengthy transcript into a concise summary highlighting the main discussion points, decisions, and outcomes.

   - *Prompt Example*: "Summarize this meeting transcript from our weekly sales strategy session. Focus on the key topics discussed, major decisions reached, and any unresolved issues or questions raised."
   - *Examples of background information to attach/provide*: The full meeting transcript text. Optionally, the meeting agenda or a brief statement of the meeting's purpose for better context and a list of attendees and their roles, if unclear from the transcript.

2. **Action Item Extraction and Assignment:**
   Have an AI model scan the transcript to identify all tasks, assignees, and deadlines mentioned.

  - *Prompt Example*: "Review the attached meeting transcript and extract all action items. Identify the person responsible (owner) and any specified deadline for each action item. List them in a table format: | Action Item | Owner | Deadline | Relevant Transcript Snippet |."
  - *Examples of background information to attach/provide*: The full meeting transcript text. Optionally, a list of attendees to help clarify names if the transcript is unclear.

3. **Sentiment Analysis of Key Topics or Speakers:**

   Ask an AI model to analyze the tone and sentiment expressed regarding specific subjects or by particular participants.

   - *Prompt Example*: "Analyze the sentiment expressed in this transcript regarding the new marketing campaign proposal. Identify statements indicating positive, negative, or neutral sentiments from different speakers. Provide quotes to support the analysis."
   - *Examples of background information to attach/provide*: The full meeting transcript text and the specific topic or speakers you want the sentiment analysis to focus on.

4. **Key Takeaways and Next Steps Identification:**

   Use an AI model to distill the discussion's most important conclusions and agreed-upon next steps.

   - *Prompt Example*: "Based on this project update meeting transcript, generate a bulleted list of the top three to five key takeaways and a separate list of all agreed-upon next steps for the team."

- *Examples of background information to attach/provide*: The full meeting transcript text. Optionally, the project name or meeting objective.

5. **Participant Contribution Analysis:**

   Have an AI model identify who spoke the most or who contributed key ideas or decisions on specific topics.

   - *Prompt Example*: "Analyze this transcript from our brainstorming session on guest experience improvements. Identify the participants who contributed the most distinct ideas. Also, list the key suggestions made by [Speaker Name]."
   - *Examples of background information to attach/provide*: The full meeting transcript text and a list of meeting attendees if speaker identification is ambiguous in the transcript.

6. **Identifying Questions and Unanswered Points:**

   Ask an AI model to extract all questions posed during the meeting and identify whether they were answered or remain outstanding.

   - *Prompt Example*: "Go through this Q&A session transcript. List all distinct questions asked. For each question, note who asked it and to whom it was directed (if specified) and provide a summary of the answer given in the transcript. If a question appears unanswered, please note that."
   - *Examples of background information to attach/provide*: The full meeting transcript text.

7. **Topic-Based Segmentation of Discussion:**

   Use an AI model to break down a long transcript into sections according to the main topics discussed.

- *Prompt Example*: "Segment this department meeting transcript into sections based on the primary topics discussed. Provide a heading for each topic and a summary of the discussion points under that topic."
- *Examples of background information to attach/provide*: The full meeting transcript text. Optionally, the meeting agenda, which can help guide topic identification.

8. **Extraction of Commitments and Promises:**

   Have an AI model identify any commitments, promises, or undertakings made by participants during the meeting.

   - *Prompt Example*: "Review this client negotiation meeting transcript and extract all statements where either our team members or the client representatives made a commitment or promise. Note who made the commitment and what was committed to."
   - *Examples of background information to attach/provide*: The full meeting transcript text and a list of attendees clearly identifying "our team" and "client representatives."

9. **Identifying Points of Agreement and Disagreement:**

   Ask an AI model to highlight key areas where consensus was reached and where differing opinions or disagreements were evident.

   - *Prompt Example*: "Analyze this strategic planning transcript. Identify the key proposals or topics where there was clear agreement among participants. Also, highlight any significant disagreement or debate points, noting the differing perspectives if possible."
   - *Examples of background information to attach/provide*: The full meeting transcript text.

10. **Drafting a Follow-Up Communication According to Transcript:**

    Use an AI model to generate a draft email or internal memo summarizing the meeting informed by the transcript, ready for you to edit and send.

    - *Prompt Example*: "Based on the attached transcript from yesterday's project kick-off meeting, draft a follow-up email to all attendees. The email should briefly summarize the meeting's purpose, reiterate the key decisions made, list the assigned action items with owners and deadlines, and confirm the next meeting date if mentioned."
    - *Examples of background information to attach/provide*: The full meeting transcript text, a list of attendee email addresses or a placeholder to insert them, and the desired tone for the follow-up email, such as formal, informal, or concise.

The most practical way to leverage AI with your standard office software is through a simple, manual transfer. You first use a general AI model to generate a starting point for your task, whether it's content for a Word document, a framework for a spreadsheet, an outline for a presentation, or a summary of a transcribed meeting. While this method requires you to copy and paste between applications, it provides a substantial head start on work, allowing you to focus your expertise on high-value refinement rather than on blank-page creation.

Key Takeaways:

- General AI models act as on-demand assistants for a wide array of tasks typically performed in standard office software like word processors, spreadsheets, email clients, and presentation tools, as well as for analyzing outputs, such as meeting transcripts. You do not need to wait for AI to be embedded in every application to benefit.

- The core process for leveraging external AI effectively involves defining your need, crafting a detailed prompt, providing essential background information, allowing the AI to generate output, and then implementing and refining that output within your familiar office applications. This changes manual effort into AI-assisted creation.
- The skill of detailed prompting with good background information, illustrated through our hospitality-focused examples in this chapter, allows for broad applicability across virtually any hotel department, from guest services and F&B to sales, marketing, and operations, generating valuable content, frameworks, and ideas.
- True advantage emerges when you skillfully integrate AI-generated outputs into existing workflows and applications, which requires developing strong prompting skills, applying critical human oversight for quality and accuracy, and committing to ongoing learning about AI capabilities.

Bringing This into Focus:

- **Tackle One Drafting or Planning Task with AI Assistance This Week:**
    - *Focus*: Identify one task where you typically spend significant time drafting content from scratch, for example, a complex guest response email, an outline for a new SOP, or the structure for a sales proposal.
    - *Action*: Craft a detailed prompt for this task using a general AI model, providing thorough background information as demonstrated in this guide's examples. Generate a first draft or an outline. Then, bring this AI-generated output into your preferred word

processor, email client, or presentation software and refine it with your expertise. Note the time saved or the improvement in your starting point.

- **Conceptualize a Solution or Improvement with an AI Brainstorming Partner:**
  - *Focus*: Think about a persistent challenge, a desired operational improvement, or a new initiative for your hotel, such as enhancing pre-arrival guest communication, analyzing guest feedback trends from various sources, or developing a new local tour package.
  - *Action*: Use a general AI model to brainstorm potential solutions, content ideas, data analysis approaches, or communication strategies related to this. Ask the AI to play different roles, for instance, "Act as a marketing expert and suggest three ways to promote our new F&B offering to in-house guests." Use the AI's output to create a presentation to frame your internal discussions or to develop a more concrete plan.
- **Streamline Information Processing from Meeting Transcripts:**
  - *Focus*: Summarize and review your next team meeting, which will be recorded and transcribed using tools like Zoom, Microsoft Teams, or Google Meet.
  - *Action*: Take the raw transcript and use a general AI model with specific prompts, like those shown in the "Leveraging AI with Meeting Transcripts" section of this guide, to quickly extract action items, generate a concise summary, or identify key decisions. Compare the time and effort to your usual method of processing meeting notes and share the AI-assisted summary with attendees.

Leadership Action Agenda:

- **External AI Prompting Skills and Application Initiative:**
    - *Action*: Convene department heads and potential AI champions to review the examples in this guide or identify other high-impact tasks within their teams. Pinpoint three to five key tasks per department where using a general AI model for initial drafting, brainstorming, data structuring, or transcript analysis could yield significant time savings or quality improvements.
    - *Develop*: Launch an internal awareness campaign such as Prompt Power Hour or AI-Assisted Wins of the Week and offer practical micro-sessions focused on crafting effective prompts for common hospitality tasks. Emphasize the importance of providing background information to the AI and then refining the AI's output. Encourage the sharing of successful prompt examples.

- **AI-Assisted Content and Framework Generation for a Key Process:**
    - *Action*: Select one recurring hotel process that involves substantial content creation, information synthesis, or planning, for example, developing customized event proposals, creating training materials for new seasonal offerings, or summarizing monthly departmental reports for executive review.
    - *Challenge*: Assign a small, cross-functional team to map the existing workflow. Task them with identifying specific points where outputs from a general AI model, based on detailed prompts and relevant background information, could be manually

integrated to provide better starting points for staff, ensure consistency, or accelerate the process. Have them present a revised workflow concept detailing the AI prompting stages, the expected AI output, and how this output is utilized within your existing software tools.

- **Fostering a Prompt-First, Human-Refine AI Culture:**
    - *Action*: As a leader, actively encourage your teams to view general AI models as valuable first-draft assistants, brainstorming partners, or data summarizers before they tackle tasks from a blank slate. The goal is to use external AI to overcome initial hurdles and generate a solid foundation, then apply human expertise for critical thinking, adaptation, and finalization within your standard office tools.
    - *Lead by Example*: Visibly share an instance where you have used a prompt with a general AI model to generate a useful starting point for a document, a strategic outline, a complex email, or a meeting summary, and explain how you then refined that output. Encourage team members to share their prompt success stories and AI-assisted efficiency gains in team meetings or on an internal collaboration channel, promoting a culture of smart, practical AI utilization with the tools available today.

7
CRAFTING SIGHTS AND SOUNDS

IN THE PREVIOUS chapter, we explored how general artificial intelligence models can serve as powerful external assistants, helping you generate and refine text-based content for your everyday office software. We saw how effective prompting can turn these AI into valuable partners for drafting documents, conceptualizing spreadsheet analyses, planning communications, and even dissecting meeting transcripts. Now, we venture beyond text into the rapidly developing field of AI-powered visual and auditory content creation.

The hospitality industry, at its heart, involves creating memorable experiences. The ambiance of your lobby, the appeal of your marketing materials, the clarity of your guest communications, and even the hold music on your phone system all contribute to the overall brand perception. As we examine the concepts, you will discover how AI is a creative partner for crafting the very sights and sounds that define your hotel's unique identity and guest experience. This chapter will guide you through understanding and utilizing AI for image, video, and even voice generation, significantly changing how you approach branding, marketing, and communication.

AI in Visual Content Generation

The ability to generate unique, high-quality images from simple text descriptions, often called text-to-image generation, is changing visual content creation. Instead of relying solely on stock photography or lengthy design processes, hospitality professionals can now use AI to bring their specific visions to life. This technology is accessible through a growing array of applications and platforms. Key players in this space include DALL-E 3, Midjourney, Flux, Imagen 3, and Ideogram, alongside platforms that integrate these capabilities, such as Canva and Adobe Firefly.

What makes these new AI image generators so powerful? Let's explore some of their enhanced features.

Understanding Modern AI Image Generation Capabilities

The latest generation of AI image tools offer significant advancements that address many of the limitations of earlier systems. Understanding these capabilities is key to leveraging their full potential.

- **Native Multimodality**: Modern AI image generators are increasingly designed with native multimodality. This means they are built from the ground up to understand the intricate relationship between textual descriptions and visual elements. They connect the words you use in your prompt to visual concepts, styles, and compositions.
- **Near-Perfect Text Rendering**: A significant leap forward is the improved ability of AI to render legible and contextually accurate text within images. Historically, getting AI to spell correctly or place text naturally on an image was

a major challenge due to models focusing on pixels. Some new tools employ an autoregressive approach, generating images sequentially, for example, from top left to bottom right rather than all at once, which contributes to better text rendering and overall image coherence. This opens new use cases for information-dense visuals.

- **Multi-Turn Generation and Context Maintenance**: One of the most user-friendly advancements is the ability to engage in a "conversation" with the AI to refine an image. The AI can maintain context from previous prompts and images within the same session. If you generate an image and then say, "Make the cat fluffier" or "Change the sky to sunset," the AI understands you are referring to the existing image. Users can even ask the AI for suggestions on how to improve an image, receiving prompts for enhancing backgrounds, adding elements, refining characters, or changing the lighting.
- **World Knowledge Integration**: Because these models are often trained on vast datasets containing both text and images simultaneously, they possess a deep world knowledge about visual concepts, styles, and real-world objects. This allows them to generate images that are more than just visually appealing but also contextually relevant.
- **"Binding:" Getting Attributes Right:** Binding in AI image generation refers to how well the model maintains correct relationships between attributes like colors or textures and the objects they are supposed to describe. A model with poor binding might confuse instructions, for example, creating a red star when asked for a blue star and a red triangle. Newer tools show significant improvement, sometimes correctly binding attributes for fifteen to twenty objects without confusion. Still, challenges remain, especially when rendering more than fifteen or twenty distinct concepts or objects simultaneously or with non-Latin text fonts.

- **Character Consistency (The Ideal and the Reality):** The goal of character consistency is to maintain the appearance of a specific character across multiple images or iterations. Although this is a key area of development, it can still be challenging. Sometimes, the character changes subtly with each new prompt or refinement, akin to a game of telephone, where the message distorts with each retelling.
- **Image Description and Re-creation:** An impressive capability is the AI's ability to look at an existing image and provide a detailed textual description of its elements, including characters, background, details, lighting, and mood. You can then use that AI-generated description as a prompt to re-create the image. This can be a great way to understand how the AI "sees" an image and to generate variations with a similar essence, although each new version might still have slight changes from the original.

Practical Applications for Hospitality Visuals: Bringing Your Brand to Life

These AI capabilities open a world of possibilities for the hospitality industry. Let's explore some practical ways you can use AI image generation to enhance your brand and operations.

- **Create Your Visual Brand System**: Go beyond traditional design processes to rapidly prototype and visualize core branding elements.
    - *Use Cases*: Generate logo variations, visualize color palettes in different applications, create social media post templates, design website header images, or produce product mockups, such as branded amenities.

- **Visualize Complex Concepts and Data:** Make intricate information more accessible and engaging through AI-generated visuals.
    - *Use Cases*: Develop clear process diagrams, such as for internal training on a new guest check-in procedure; create concept maps to explain service strategies; design comparison charts for different room packages or event spaces; produce step-by-step visual guides for guests, like how to use in-room technology; or craft metaphorical illustrations of abstract brand values.
- **Scale Your Content Production with a Consistent Visual Style**: Once you have defined a visual style, AI can help you efficiently apply it across a wide range of content.
    - *Method*: Create a visual style guide for your content, for example, "minimalist illustrations with a primary color palette of blues and greens, and a slight grain texture, suitable for a serene spa brand." Generate a template image that perfectly captures this style. For each new piece of content, reference that template image if the tool allows image input or re-use the successful prompt elements, asking for variations that maintain the style and change the subject matter.
- **Create High-Converting Marketing Assets**: Develop compelling visuals designed to capture attention and drive guest action.
    - *Use Cases*: Generate before/after visualizations, such as a recently renovated room or amenity; design illustrative customer journey maps for marketing presentations; create product/service benefit

illustrations, for instance, visually showing the relaxation achieved from a spa treatment; produce aspirational lifestyle imagery that speaks directly to your ideal guest demographic; and even create testimonial visuals that incorporate snippets of real guest feedback in an engaging way.

Mastering the Art of Image Prompting: Your Dialogue with AI

The quality of the images you generate is directly proportional to the quality of your prompts. Think of prompting as giving creative direction to an artist; the more clearly and comprehensively you can articulate your vision, the closer the AI will get to producing it.

The specificity principle is essential. The more specific your prompt, the better the AI can generate an image that matches your expectations. If you want a complex image, describe the scene thoroughly by detailing the main subjects, their actions, the setting, and even background elements. Use clear, unambiguous language, avoiding jargon or words with conflicting meanings that might confuse the AI; for instance, using "imaginary" and "realistic" together could be problematic. Leverage adjectives to define your image elements as closely as possible.

Which prompting approach should I use for AI image generation?

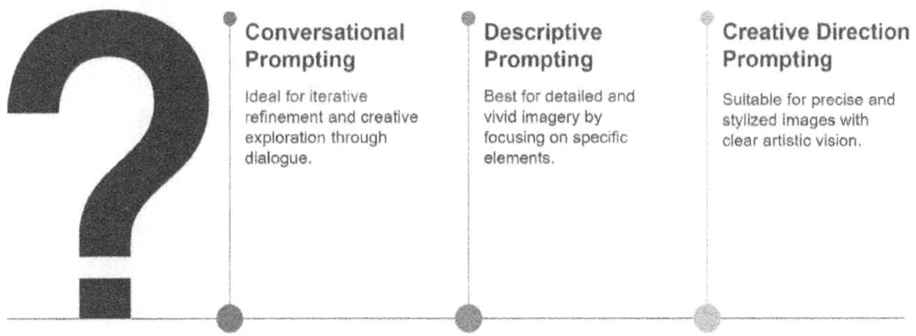

Three key prompting approaches to "talk" to the AI image generator.

1. *Conversational Prompting:* This allows you to engage with the AI as you would a creative friend. Start with a general idea and then refine it through iterative dialogue, adding more detail with each step.

   Hotel Example Dialogue:

   - **You (Prompt 1):** "Create an image for our hotel's new Summer Seaside Escape package."
   - **AI:** *(Generates an initial image, perhaps of a generic beach scene.)*
   - **You (Prompt 2):** "That's a nice start. Can you add a luxury hotel balcony overlooking the scene? Make the ocean a vibrant turquoise color."
   - **AI:** *(Generates a new image with a balcony and different watercolor.)*
   - **You (Prompt 3):** "Perfect. Now, place a small table on the balcony with a breakfast tray for two, including fresh orange juice and croissants."
   - **AI:** *(Adds the breakfast tray to the image.)*
   - **You (Prompt 4):** "Excellent. For the final touch, make it early morning with soft, golden sunlight."

2. *Descriptive Prompting:* This involves imagining you are describing the desired image to someone who cannot see it. Focus on capturing details like colors, emotions, the subject, setting, mood, and artistic style. A simple structure for this is [Subject] + [Setting] + [Mood] + [Style or Color].

   Hotel Prompt Example:

   "A family of four laughing on a sun-drenched, pristine hotel beach during a bright afternoon. The mood is joyful, relaxed, and full of warmth. The image should be in a vibrant, photorealistic style with bright, natural colors."

3. *Creative Direction Prompting:* This means treating the AI as your skilled graphic artist. You provide clear vision, specifying elements like shot composition, lighting, emotion, artistic style, genre, and camera angle. A prompt structure here could be [Subject] + [Environment/Scene] + [Lighting] + [Mood] + [Camera Angle] + [Art Style or Medium].

   Hotel Prompt Example:

   "A close-up shot of a hotel guest's hands receiving a room key card from a smiling concierge at a polished wooden front desk. The scene is the hotel lobby, which is modern and filled with lush green plants in the softly blurred background. Use warm, welcoming, and bright indoor lighting. The mood is efficient, upscale, and hospitable. The camera angle should be eye-level with a shallow depth of field. The final image must be in the style of high-quality commercial photography."

Avoid expecting perfection on the first try. The true strength of AI image generation lies in the iterative process. Start general, then refine; experiment with different phrasings, as AI can interpret prompts unexpectedly. Use reference anchoring in your prompt if you admire a particular visual style. If you are struggling, you can even use a chatbot for prompt assistance by describing what you want and asking the chatbot to generate a detailed prompt for an image.

Editing AI-Generated Images

Once an image is generated, some AI tools offer capabilities to edit it, although this area is still developing. For example, within the ChatGPT interface, users can often select parts of an image or provide further instructions to modify it. Microsoft Designer also provides more comprehensive design tools for refining AI-generated visuals. Still, precise, targeted editing without affecting other elements can be challenging, and some tools may have bugs or inconsistencies, especially with complex edits like faces over multiple passes.

Navigating Limitations and Safeguards

Although incredibly powerful, AI image-generation tools have certain limitations and necessary safeguards. Common limitations include cropping issues, where images may sometimes get cropped, especially near the bottom. Hallucinations can occur, where image generators, like all AI systems, occasionally "hallucinate" or invent information or details, particularly if the prompt is too vague or lacks context. Concept overload is another issue; asking the AI to include too many distinct concepts in a single image can lead to confusion or inaccurate rendering. Rendering text accurately in languages that do not use the Latin alphabet, such as Japanese or Arabic, can be inconsistent. As mentioned, making very specific, isolated edits to an image without unintended changes to other areas remains a hurdle. Also, be prepared for images to take some time to generate, typically around thirty to sixty seconds, depending on the complexity and the specific tool.

AI providers implement safeguards and content policies to prevent misuse. These include preventing watermark removal from generated images and blocking the generation of harmful content, such as sexual deepfakes or child sexual abuse material (CSAM). AI image generators are generally programmed to avoid reproducing copyrighted characters due to copyright restrictions; for example, a request to "create an image of Mickey Mouse" will likely be refused. The AI may offer to create an original character in a similar style instead. Requests that violate content policies will often result in a failure to generate the image, sometimes with a message explaining the violation. For example, a prompt was refused because it involved manipulating images of real people in ways they did not explicitly consent to. Even blurring faces might not bypass such policies if the underlying intent is still problematic.

AI in Video and Voice Generation

The principles of AI-driven content creation extend beyond static images into the dynamic fields of video and audio.

- AI video generation with Sora or Veo 3 allows you to provide a text prompt, and the AI generates a video clip. The techniques often build upon those used for image generation and incorporate elements of motion and temporality, requiring prompts to include action-related words and descriptions of movement. Prompting for video is like image generation; you can use long, detailed prompts or shorter, concise ones, and trial and error are still part of the creative process. You can even use a text-based AI to help you brainstorm and craft detailed video prompts by describing the scene, characters, actions, and desired cinematic style.

  Current constraints and considerations for AI video generation include limited availability, as access to cutting-edge models like Sora might be restricted or not initially included in standard team subscriptions. Videos generated by tools like Sora or Veo 3 may be watermarked by default. To prevent misuse, AI providers employ content moderation safety measures, including prompt filtering and output classifiers. Subscriptions might have specific usage limits on the number of videos, resolution, or length. Video generation is complex, and current models can sometimes produce errors like objects unnaturally disappearing, duplicating, or "melting." This field is quickly developing, so features, limitations, and access will likely change over time.
- AI voice generation involves AI synthesizing human-like speech from text (text-to-speech) or cloning existing voices

with ethical considerations and necessary permissions. These include ElevenLabs and Murf AI for voice generation and Suno for AI-generated music. Hospitality applications for this capability are numerous. Imagine using AI-generated voices for narration for staff training videos in multiple languages, professional and consistent voiceovers for virtual tours of your property, personalized welcome messages for in-room smart devices, dynamic announcements or informational updates, or creating unique, royalty-free background music or soundscapes for different areas of your hotel.

- AI music generation is another rapidly advancing capability, with platforms like Suno enabling the creation of original musical pieces through AI. Typically, users provide text prompts that describe the desired genre, mood, tempo, instrumentation, or even lyrical themes, and the AI then composes a unique track.
- AI avatars are digital human-like figures that AI can create and animate to speak and gesture. Platforms like HeyGen and Synthesia specialize in creating AI avatars. Although still an emerging area, AI avatars could eventually be used in hospitality for interactive virtual concierges on your website or in-lobby kiosks, digital hosts for online events or webinars, or engaging characters in interactive training modules.

Adopting the AI Mindset for Visual and Auditory Creation in Your Hotel

Embracing these AI tools for sights and sounds requires cultivating an AI mindset across your team. This means recognizing that AI is an augmenter of human creativity. The goal is to create a coordinated ecosystem with AI tools, whether they are the general AI models assisting with your office application tasks or these specialized tools for image, video, and voice. This synergy can lead to more seamless workflows, richer and more personalized guest communications, and a more innovative and dynamic brand presence.

Still, this journey requires thoughtful human guidance. Effective prompting and human oversight are essential to ensure quality, brand alignment, and ethical use, for example, respecting copyright and avoiding harmful misrepresentations. As these technologies continue to develop, a commitment to ongoing learning and adaptation will be essential for any hospitality business looking to truly leverage the power of AI in crafting its unique sights and sounds. Consider this chapter your invitation to start painting with words and composing with algorithms. The creative potential is immense, and the ability to quickly and cost-effectively generate unique visual and auditory assets for the hospitality industry can be a major advancement in how you connect with guests and tell your brand's story.

Coordinating Your Brand's AI-Powered Identity

The ability to translate textual ideas into visuals or audio using AI tools marks a huge shift. You are now an active collaborator with a powerful creative partner. The key to this partnership exists in understanding the tools, mastering the art of prompting, and embracing an iterative process of creation and refinement.

Key Takeaways:

- AI has evolved into a multimedia creative force, enabling the generation of a wide spectrum of visual content (from logos and marketing assets to complex data visualizations) and auditory elements (like voiceovers and music) directly from textual prompts, expanding beyond text-based assistance.
- Modern AI image-generation tools boast advanced capabilities, such as understanding the relationship between text

and visuals (native multimodality), improved rendering of text within images, multi-turn conversational refinement, and better attribute binding. These new capabilities provide significant potential for creating custom visuals for dynamic marketing, training materials, and unique branding in hospitality.

- Effective prompting is your creative north star for AI-generated sights and sounds. Mastering various techniques (conversational, descriptive, creative direction) and utilizing specific attributes is vital for guiding AI. Embracing iteration and exercising human oversight ensures brand alignment, quality, and ethical use.
- The principles of AI-driven content creation now extend to AI video and voice generation. There are tools that can produce video clips from text and synthesize human-like speech or music, offering new avenues for hospitality applications. These include training narrations, virtual tours, and personalized guest messages.

Bringing This into Focus:

Here is how you can start:

- **Generate Your First AI Visuals This Week:**
    - *Task*: Choose a simple, tangible concept relevant to your hotel, perhaps a welcoming image of your lobby from a unique perspective, a visual for a daily special at your restaurant, or an icon representing a guest amenity.
    - *Action*: Try generating this visual using at least two of the prompting styles discussed:
        - Descriptive Prompting: Write a detailed paragraph describing the scene, colors, mood,

and desired style, for example, "A photorealistic image of a cozy hotel lobby armchair next to a fireplace, warm lighting, inviting mood, rich textures."

- Conversational and Iterative Prompting: Start with a simpler prompt like "Hotel lobby armchair by the fireplace" and then add details or request modifications in subsequent prompts, such as, "Make the lighting warmer," "Add a book on the side table," or "Change the style to watercolor illustration."

- *Reflect*: Compare the results. Which prompting style felt more intuitive? How did iteration change the outcome?

- **Visualize a Piece of Hotel Information or Process:**

  - *Focus*: Select a piece of information or a simple process within your hotel that you often explain using only text, for instance, the three key steps for guests to access hotel Wi-Fi, a visual comparison of two different tour packages offered by the concierge, or a simple "Do Not Disturb" sign concept with unique branding.
  - *Action*: Attempt to create a visual aid for this using an AI image generator. If your chosen tool has good text rendering capabilities, try incorporating short, clear text directly into the image. Focus on clarity and immediate understanding. For instance, prompt for "A clean, minimalist infographic showing three simple steps with icons for connecting to hotel Wi-Fi, using our hotel's color palette."

- **Conceptualize a Multimedia Marketing Snippet:**
    - *Focus*: Think about a short promotional message for your hotel, such as announcing a new seasonal menu, highlighting a weekend spa offer, or a book direct incentive.
    - *Action*:
        - First, use an AI image generator to create a compelling visual for this promotion.
        - Next, if you have access to or are exploring AI voice generation tools, like ElevenLabs or Murf AI, draft a short (fifteen to thirty second) script for this promotion and hear how it might sound with an AI-generated voice.
        - Conceptually, how could these two elements, an AI image plus an AI voice, combine, perhaps even as a precursor to thinking about AI video generation like with Sora or Veo? This exercise is about understanding the potential synergy.

Leadership Action Agenda:

Leading your hospitality organization into this new era of AI-driven content creation requires vision, encouragement, and a framework for responsible innovation.

- **Launch a Creative AI Sandbox Initiative:**
    - *Action*: Designate a small, cross-functional team, for example, from marketing, sales, F&B, and guest services, to specifically explore and experiment with

AI image, voice, and, if feasible, video generation tools for a defined period, perhaps one month.

  - *Focus*: Task them with tackling two to three real-world content needs for the hotel, such as creating a new set of social media visuals for an upcoming holiday, developing concept art for a proposed themed event, or producing a draft AI-narrated virtual tour script. Have them document their prompting strategies, iterative processes, tool experiences (both successes and challenges), and the perceived impact on efficiency and creativity.

- **Develop AI-Assisted Brand Expression Guidelines:**

  - *Action*: As your teams begin to explore AI content generation, initiate a process to develop internal guidelines for its use. This ensures brand consistency and ethical application.
  - *Considerations*: How can AI-generated visuals align with your established brand aesthetics, including colors, style, and tone? What are the standards for using AI-generated voices to represent the brand? What are the internal review processes for AI-generated content before it is used externally? How will you address issues of copyright, authenticity, and potential bias in AI outputs?

- **Champion a Culture of Co-Creation with AI:**

  - *Action*: Actively promote the mindset that AI tools are collaborators. Encourage your teams to see themselves as creative directors, guiding the AI and embracing the iterative process of refinement that leads to the best results.

- *Showcase and Share*: Create opportunities for employees to share their AI-generated creations, successful prompts, and learnings, perhaps in team meetings or an internal AI Innovation Hub. Visibly support and reward creative experimentation and problem-solving using these tools. Emphasize that the goal is to augment human creativity and free up time for higher-value strategic and guest-facing work, aligning with the AI mindset principles of achieving greater productivity and work enjoyment.

PART 3

RESPONSIBLE AND EFFECTIVE IMPLEMENTATION

8
ETHICS, RESPONSIBILITY, AI GOVERNANCE

Guiding AI Principles

AS WE INTEGRATE AI more deeply into our operations, it is imperative we shift our focus from simply what AI can do to how we ensure it is used wisely. This chapter is dedicated to the essential principles that must guide our AI journey: responsibility, ethics, and safety. These are the practical foundations for building trust and ensuring AI serves us, our teams, and our guests well.

Let's start by understanding what each of these terms means in the context of AI.

- **Ethical AI**: This serves as your moral guide, addressing the values and principles that steer AI's design, development, and application. It fundamentally questions *why* AI is used and ensures its alignment with core hotel values, such as guest respect, team well-being, fairness, inclusivity, and empathy. For instance, with guest personalization, ethically trained staff must define the boundary between helpfulness and intrusion to maintain trust. Similarly, human ethical oversight in HR applications is essential to confirm that AI supports fairness and employee development rather than enabling impersonal surveillance. Decisions about disclosing AI's non-human nature to guests also fall under human ethical judgment. Ultimately, ethical AI provides a human-driven critical perspective for assessing AI's effect on human values.

- **Responsible AI**: This focuses on *how* AI is integrated accountably and transparently within your operations. It encompasses the concrete practices, policies, and human oversight necessary to ensure AI respects legal standards and upholds brand promises, always with a human in the loop at key junctures. Such practices include AI governance committee approval for new AI tools, human assessment of AI training data for bias, and human validation of new AI features before they impact guests or influence high-stakes decisions. Clear internal AI usage rules enforced by human managers and processes for skilled human review and override of AI suggestions for critical decisions are also essential to maintain control. Documenting AI limitations and establishing clear channels for staff to report concerns are further components of this approach. This responsibility is embedded in tangible, human-managed operational structures.
- **Safe AI**: Safe AI addresses the critical question, "What if things go wrong?" by focusing on preventing or mitigating potential harm from AI systems to guests, staff, your brand, or finances. Key safety considerations involve ensuring the AI reliably performs its intended functions without causing unintended harm, such as discriminatory pricing or data exposure. It also means confirming that the AI operates

safely for staff, for example, by avoiding unfair evaluations and protecting the brand from misinformation or misuse. A significant risk is AI hallucinating dangerously incorrect information, like wrong emergency procedures. An AI system is deemed unsafe if its real-world operation yields systematically biased or critically inaccurate outputs, regardless of how ethically it was designed. Therefore, continuous human monitoring and the capacity for swift human-in-the-loop intervention are fundamental to AI safety. Without this human-verified assurance of operational safety, even well-intentioned ethical designs and responsible processes can be compromised.

These three lenses—ethical, responsible, and safe AI—work together with the human in the loop as an active, essential component, ensuring AI contributes positively and safely to your hotel.

The Human in the Loop Cycle

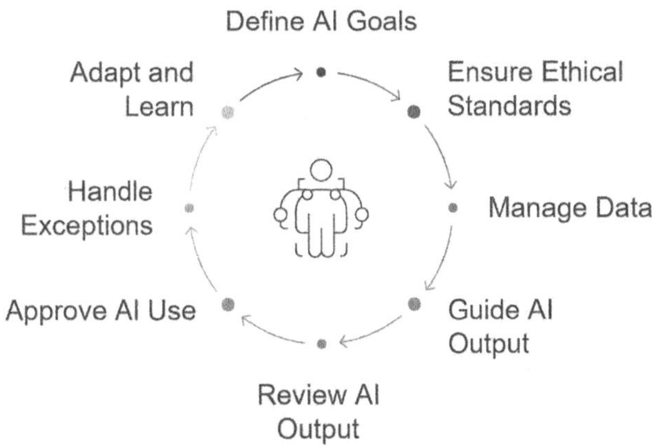

The Human in the Loop (HITL)

- Deploying artificial intelligence effectively and ethically in hospitality fundamentally depends on continuous, intelligent human participation. The human-in-the-loop (HITL) principle means that humans guide, manage, and ensure AI responsibly serves its intended purpose. In this vital HITL capacity, humans act in several key roles. I am calling out the following functions to emphasize them as distinct *human* responsibilities; while aspects of these tasks might appear automatable, assigning them to a person is a deliberate reminder that these vital duties should not be fully offloaded to AI.
- **Strategic Directors**: Defining AI goals, intended uses, and overarching ethical boundaries.
- **Ethical Overseers**: Ensuring fairness, transparency, accountability, and guest trust in all AI applications.
- **Data Curators and Trainers**: Selecting, cleaning, and managing data used to train custom AIs and providing ongoing feedback to refine AI performance.
- **Prompt Engineers and Content Guides**: Instructing AI to produce desired outputs and align with brand voice.
- **Critical Reviewers and Validators**: Carefully checking AI outputs for accuracy, relevance, bias, and potential harm before use.
- **Final Approvers**: Authorizing the use of AI-generated content or AI-informed decisions, especially in high-stakes or guest-facing situations.
- **Exception Handlers**: Managing scenarios, detailed situations, or complex emotional contexts that fall outside AI's designed capabilities.
- **Continuous Learners and Adapters**: Staying informed about AI development and refining HITL processes as technology and understanding mature.

The Ethical Situation of AI in Hospitality: Key Issues and Human Oversight

To use AI effectively in hospitality, we must understand specific ethical issues and how HITL helps manage them.

- **Algorithmic Bias:**

  AI systems learn from the data they are trained on. If this data contains existing societal biases, the AI can unintentionally continue or even amplify these biases. Bias can affect hiring by disadvantaging candidates, pricing by setting unfair rates based on profiles, service personalization by favoring groups, marketing by reinforcing stereotypes, and feedback analysis by skewed weighting.

  - *Proactive Mitigation:*

    - Question AI vendors about their models' training data sources and diversity efforts.
    - For custom AIs, curate varied and representative internal data for training to avoid skewed outcomes.
    - Conduct regular bias audits by reviewing AI outputs for unfair results across different guest and employee groups.
    - Ensure skilled human oversight is in place for important decisions, with the ability to override AI suggestions.
    - Train your team to spot and question potential AI biases in their daily work.

- **AI Hallucinations and Misinformation—When AI Invents Information:**

  Generative AI does not know facts in a human sense; it predicts language. It can create false information with complete confidence. An AI chatbot giving incorrect pool hours causes guests frustration. An AI inventing local restaurants damages credibility. AI creating wrong safety procedures is dangerous.

- *Proactive Mitigation:*
    - Mandate the cross-referencing of all AI-generated facts with official, verified hotel sources before any use.
    - For guest-facing AIs, ground their knowledge in a curated and verified library of your hotel's accurate information.
    - Implement a clear HITL review and approval process for all critical guest communications or internal documents drafted with AI.

- **Intellectual Property, Copyright, and Legal Considerations:**

  When AI helps create copy, visuals, or music, questions of ownership arise. AI output comes from patterns in its training data, which may include copyrighted material. Using an AI-generated image that is too similar to copyrighted work can lead to legal problems. If AI drafts your unique service manual, your hotel's claim to it as your intellectual property may be complex. This legal area is still developing.

    - *Proactive Mitigation:*
        - Keep detailed records of how and where AI is used in the content creation process.
        - For core IP like brand slogans or signature service methods, ensure substantial human creative direction and original changes are made to any AI-assisted draft.

        - Use AI primarily for initial ideas for important brand assets, like logos, with final creation handled by human designers.
        - Regularly consult with legal experts specializing in IP and AI law.

Beyond intellectual property, it is necessary for hoteliers to recognize that the legal environment for AI is quickly developing. Your AI governance committee, in consultation with legal experts, should consider liability, as complex questions arise if an AI system's error leads to guest harm or financial loss. Employment law is another consideration, as using AI in recruitment or performance monitoring requires human oversight and legal review to ensure fairness. Adherence to data privacy regulations like GDPR and CCPA is essential to avoid significant penalties, and you must also account for any industry-specific regulations relevant to your hotel's offerings. Establishing a relationship with legal counsel knowledgeable in AI law is necessary for responsible AI adoption.

- **Data Privacy and Security: Protecting Guest Trust**

  Hotels are custodians of highly sensitive guest information, from personal identifiers and payment details (PII, PCI) to stay preferences. The integration of AI significantly elevates the importance of protecting this information, as a security failure can lead to severe reputational damage and legal penalties. AI systems can introduce new vulnerabilities, such as data breaches through the tools themselves, the inadvertent feeding of sensitive data into public models by untrained staff, or risks associated with the practices of third-party vendors.

    - *Proactive Mitigation:*

        - **Enforce Strict AI Governance:** Your AI governance policy must clearly list all approved AI tools and the data types they can be used with. Enforce a non-negotiable rule that no PII, PCI, or confidential hotel information is ever entered into unapproved public AI tools.

  - **Mandate Secure, Enterprise-Grade AI:** For any process involving sensitive data, select enterprise-level AI platforms that offer robust encryption, security certifications (e.g., SOC 2, ISO 27001), and contractual assurances that your hotel's data will not be used to train public models.

- **Sustainability and Environmental Impact: AI's Energy Use**

  Large AI models use a lot of electricity for training and daily operations. As our industry focuses on sustainability, AI's energy use adds to our responsibilities. Although AI can help optimize energy use in hotels, the AI systems consume significant energy.

  - *Proactive Mitigation:*

    - Encourage purposeful AI use, avoiding excessive or unnecessary querying.
    - For some back-office tasks, schedule batch AI requests to improve energy efficiency.
    - Consider using less computationally intensive (and thus more energy-efficient) AI models for less complex tasks where feasible.
    - Inquire with vendors about their use of renewable energy and sustainability practices.
    - Include AI's energy consumption in the hotel's overall environmental strategy.

- **Job Displacement and Workforce Transformation: Managing Change**

  The automation of tasks by AI naturally raises concerns about jobs. This is a key ethical consideration in your AI ascension. Roles with many repetitive tasks will see those tasks increasingly handled by AI, which can cause staff anxiety.

  - *Proactive Mitigation:*

    - Frame AI as a tool to empower employees, augmenting their roles rather than replacing them.
    - Proactively identify future skill needs and invest in reskilling and training for current employees.
    - Analyze how roles will evolve; for instance, a reservations agent might become a guest journey specialist, using AI for enhanced personalization.
    - Maintain open and transparent communication with employees about AI implementation and the evolution of their roles.

- **Transparency and Explainability in AI Interactions: Building Trust**

  Transparency about AI use and an AI's ability to explain its outputs are ethically important. Guests may feel deceived if they believe they are interacting with a human when it is an AI. It erodes trust if a decision affecting a guest or staff member cannot be explained.

  - *Proactive Mitigation:*
    - Set clear guidelines on when and how to disclose AI interactions to guests.
    - When choosing AI tools, prioritize those with explainability features.
    - Equip staff to explain the role of AI in any assisted decision and to serve as the final point of contact.

- **Accountability for AI-Driven Decisions: Humans Remain Responsible**

  When an AI is part of a decision that leads to an error, clarity on responsibility is needed. If an AI pricing error causes a revenue loss or an AI-generated response offends a guest, blaming the AI is an unacceptable solution.

  - *Proactive Mitigation:*
    - Maintain skilled human oversight and ultimate responsibility for all high-stakes decisions.
    - Clearly define roles for those who set AI parameters, monitor performance, review outputs, and intervene when needed.
    - Ensure AI systems maintain audit trails or logs to help understand how decisions or errors occurred.

A Practical Framework for Effective AI Governance

Your governance journey starts now. Start now, start simple, and continuously improve. Create baseline policies for high-priority concerns, such as data privacy, guest-facing AI, and strategic alignment, from your initial AI committee. Develop a straightforward evaluation process for new AI initiatives. As AI use matures, so will your governance. Hotels building these foundations now gain a significant advantage. Governance is about how to do AI right. It significantly changes AI into a strategic asset that enhances guest experience, empowers your team, and strengthens your competitive position. Thoughtful hoteliers who balance innovation with responsibility will win.

Successfully managing these ethical complexities requires a proactive approach to AI governance, where the HITL is a central design principle. AI governance creates clear guardrails that enable confident and responsible innovation. This structured approach saves leadership time and prevents costly errors.

Why Governance Matters: Beyond the Buzzwords

Most hotels either avoid AI or jump in without a plan; both are dangerous. Avoidance leaves you vulnerable. An unstructured approach creates inconsistent experiences, privacy violations, and conflicting AI systems. Governance significantly changes AI from an experiment into a strategic asset, giving you the confidence to move quickly without sacrificing quality or ethics. Thoughtful implementation is key. This governance typically rests on three key pillars: AI governance, AI policy, and escalation paths.

Pillar 1: AI Governance Committee

AI oversight needs a central guiding intelligence: your AI governance committee. This human committee is the strategic steering group and ethical conscience for your hotel's AI journey, balancing innovation with risk management and always ensuring that humans remain firmly in the loop of strategic AI deployment.

- Composition: Include a senior leader, representation from commercial and operational departments (perhaps their AI champions), an AI champion representative for frontline perspective, an IT/digital significant change lead, HR/L&D, and legal/compliance access. Representation requires enthusiasm, skepticism, and practical wisdom.
- Mandate and Authority: The committee should meet regularly with a structured agenda. Its core purpose is threefold:

    - *Strategic Oversight and Integration:* Reviewing and approving significant AI use cases, assessing benefits against risks, and ensuring alignment with values, strategy, and legal obligations. Humans decide the strategic fit.
    - *Enabler of Responsible Innovation:* Clearing the path for safe innovation by defining approved AI tools, use case guidelines, best practices for data handling and HITL protocols, and a pathway for teams to pilot new AI applications within a safe, human-approved framework.
    - *Responsive Issue Management and Risk Assessment:* Ensuring significant AI issues or ethical dilemmas are thoroughly investigated and effectively addressed by humans and learnings are used to refine policies and training. The committee must have real authority to pause or modify AI initiatives if concerns arise.

- Role: The committee acts as guardians of ethical standards and champions for change.

Pillar 2: AI Governance Policy

Your AI governance policy is the daily operational playbook. It must be clear, concise, and easily understandable, with just a few pages written in plain language. It needs regular review and updates by your AI governance committee, and all team members using AI must understand it. This investment of time in policy clarity is vital.

A strong AI governance policy, with HITL principles embedded, should clearly answer:

- Approved AI Tools and Usage: Which AI tools are approved for what purposes, and what are the HITL requirements for each? List sanctioned tools and specify when human review is mandatory. An example policy snippet could be: "Approved Tool: Hotel XYZ Enterprise ChatGPT. Permitted Use: Drafting initial responses to standard guest inquiries (final human review and approval mandatory before sending). Prohibited Use: Making final decisions on guest compensation without human approval."
- Data Handling and Security: How must all hotel data be handled, especially sensitive information, with clear human accountability points? Reiterate that PII, PCI, and confidential employee records should be absent from unapproved tools. Detail processes for human oversight of data handling by AI and align with data security regulations like GDPR and CCPA.
- Ethical Standards and Permitted Uses: What AI uses are permitted or encouraged, and what is prohibited, with HITL considerations? Define principles that cannot be negotiated for fairness, transparency, and disclosure. Furthermore, the policy must clearly address and prohibit the misuse of AI by employees that could be construed as cheating or a breach of workplace integrity. This includes but is not limited to misrepresenting AI-generated work by submitting it as one's original effort without proper acknowledgment or in contexts where original human work is expected; bypassing learning and assessment by using AI tools to unfairly complete mandatory training modules, competency assessments, or any evaluation designed to gauge individual knowledge or skills; inappropriate personal use by leveraging

hotel-provided AI tools and resources for significant personal projects or activities unrelated to work during work hours, especially if it incurs cost or diverts resources; and neglecting skill development through over-reliance by consistently using AI as a crutch to avoid developing or applying essential job skills, thereby failing to fulfill the expectation of professional growth and contribution. Your AI governance policy should clearly set expectations for all team members' ethical and responsible use of AI tools, emphasizing integrity, accountability, and the importance of AI as a tool to augment genuine human effort and skill, avoiding circumvention of responsibilities or compromise of workplace standards.

- Roles and Responsibilities: Who is responsible for AI oversight, policy updates, and approving new AI tools or use cases? Identify the AI governance committee and AI champions.
- Escalation and Incident Reporting: What is the process for humans to escalate concerns or report AI incidents? Detail trusted escalation pathways and reassure staff that reporting is encouraged and protected.

Bringing Governance to Life

Effective AI governance is a living framework that requires ongoing education, diligent enforcement, and a commitment to continuous development.

- Education: A policy is only as effective as the team's understanding of it. Your first step is to ensure everyone grasps the rules and the reasoning behind them. Develop tiered training programs, such as AI Essentials for All, a foundational module covering basic AI concepts, the hotel's ethical principles for AI, data privacy basics, and how to use any universally adopted AI tools responsibly. Include department-specific modules and customized training that addresses the specific AI tools, use cases, and potential risks relevant to each department, for instance, marketing's use of AI image generation or the front office's use of AI for guest communication. Provide advanced training for AI champions through in-depth sessions covering advanced prompting, ethical decision-making in AI scenarios, new AI tools being piloted, and their role in mentoring and guiding their peers.

- Enforcement: Clear guidelines require consistent application and oversight. The AI committee, or a designated governance team, should conduct regular audits of AI implementations across departments to ensure compliance with established policies and ethical guardrails. Alongside the escalation paths for urgent issues, create simple reporting mechanisms, such as a dedicated email address, a form, or a specific AI champion contact, for employees to ask questions or report general concerns about AI use without it needing to be a major alarm. Governance must have teeth; show a willingness to act by being prepared to pause, modify, or even halt AI initiatives if they drift outside agreed-upon ethical or operational guardrails or if they are found to be causing unintended negative consequences. This demonstrates a serious commitment to responsible AI.
- Development: The world of AI is anything but static; therefore, your governance framework cannot be a one-time setup. It must be a dynamic and developing system. The AI committee should schedule a regular review cadence for the entire AI governance framework: policies, training materials, champion program, and escalation paths. Quarterly is a good starting point, with the flexibility to convene more often if significant new AI developments emerge. These reviews must be informed by learning from experience within your hotel: What issues have escalated? What are AI champions reporting from their departments? Which policies are

working well, and which are proving difficult to implement or are perhaps outdated? The AI committee must also proactively monitor the external AI environment by staying abreast of external AI developments. This includes tracking emerging AI trends like new generative AI capabilities or novel applications in hospitality, significant technological breakthroughs, important regulatory announcements or discussions, and developing best practices in AI ethics and safety from the wider industry. This external vigilance allows your organization to anticipate new opportunities, foresee potential operational or ethical impacts from new AI tools or techniques, and adapt your strategies and governance framework to ensure your hotel remains both innovative and responsible in a rapidly changing environment. Based on internal experience and external insights, be prepared for iterative adaptation of your governance framework. This might mean revising policies, updating training content, introducing new guidelines for newly adopted AI tools, or refining your risk assessment processes. By embedding these practices of education, enforcement, and continuous development, your AI governance framework will become a powerful enabler of responsible innovation, guiding your hotel to harness the benefits of AI while navigating its complexities with confidence and integrity.

Making Governance Work

Bridge the gap from paper to practice by establishing clear ownership, such as an AI governance lead, to drive execution. Create a simple implementation roadmap, perhaps for three to six months, starting with quick wins. Integrate governance into existing workflows, for example, by adding an AI assessment to project approvals or making AI governance a standing agenda item in department head meetings rather than creating parallel processes.

Scaled Governance

Governance should scale. Smaller hotels might have a simplified committee and a shorter policy focusing on critical guidelines. The core principles remain, but implementation adapts. Larger groups might use a hub-and-spoke model where the corporation sets core principles, and property-level committees handle oversight.

Technology's Role in Effective Governance

Technology can support governance. Use simple tools like a centralized repository for policies, such as SharePoint or Google Drive, templates for proposals or risk assessments, or existing ticketing systems for requests or concerns. A custom GPT trained on your AI policies can help staff understand guidelines. The goal is to make compliance easier.

Common Governance Pitfalls and How to Avoid Them

Avoid making processes too rigid; teams will bypass them if they are too cumbersome. Use a tiered approach: streamlined review for low-risk uses, more scrutiny for high-risk ones. Remember that AI governance is not a one-time project. As AI develops, governance must build in regular review cycles, at a minimum quarterly. Ensure the committee has technical and operational expertise and foster knowledge sharing to avoid a knowledge gap. Do not just assume compliance; verify policies are followed through audits and safe reporting channels.

Pillar 3: AI Champions and Escalation Paths

As we navigate the integration of AI into our daily hospitality workflows, from generating guest communications and analyzing market trends to crafting marketing visuals and summarizing meeting outcomes, it is tempting to focus solely on technological capabilities. Still, technology alone will never ensure that AI is used responsibly, ethically, and safely. Your AI champions are dedicated individuals who help uphold these principles throughout the organization. They bring contextual understanding and detailed judgment that AI models inherently lack. This pillar focuses on empowering them in this role through two key strategies: cultivating AI champions and establishing clear, accessible escalation paths.

- **AI Champions**: Within every successful AI adoption, there are individuals who naturally gravitate toward understanding and effectively utilizing new technologies. These are your potential AI champions.
    - *Cultivating Your Champions*: These individuals act as internal advocates, peer mentors, and first-line support for AI-related matters within their respective departments. When identifying potential champions, look beyond mere technical proficiency or seniority. Instead, seek team members who are naturally curious and interested in learning about AI. They possess strong communication skills, enabling them to clearly explain potentially complex AI concepts; peer respect, making them trusted and approachable; and openness to new technologies, embracing change and being willing to experiment. Investing in these individuals is vital, meaning providing them with enhanced AI literacy on the principles and how to use specific AI tools. Specialized ethics training is also important, equipping them to understand potential biases in AI, data privacy implications, your hotel's specific AI usage policies, and the importance of keeping a HITL.
    - *The Champion's Role in Action*: Champions become the go-to resource for day-to-day questions, perhaps helping a colleague refine a prompt or showing how an AI tool can summarize guest preferences. They

lead by example, modeling best practices such as always cross-checking critical AI-generated information for accuracy, being transparent about AI use, and knowing when to insist that a human must be in the loop for final decisions, especially in sensitive guest interactions or staff-related matters. Because they are embedded within departments, champions often notice if an AI tool produces unexpected, biased, or ethically questionable results, such as an AI image generator creating visuals misaligned with diversity values or an AI-assisted scheduling tool showing unfair patterns. High-level AI policies are essential, and champions help bridge the gap between policy documents and daily departmental work, explaining how general rules apply to specific situations.

- **Clear Escalation Paths—Ensuring Every Voice Can Be Heard:** Although AI champions provide grassroots support, this is where clear escalation paths come into play.

    - *The Foundation: Accessible and Non-Punitive Channels*: Every single staff member, regardless of their role or technical skill, must know how to raise the alarm if they encounter unexpected AI behavior, an output that seems wrong or harmful, or a potential ethical dilemma stemming from AI use. These reporting channels must be accessible, with information on how to report a concern easy to find and the process straightforward, not buried in a dense manual. They must be non-punitive; staff must feel completely safe reporting issues without fearing blame, retribution, or being seen as anti-technology or causing trouble. The focus must always be on identifying and fixing a system or process issue, avoiding blaming the individual who spoke up. A culture that punishes reporting will quickly find itself blind to emerging AI risks.

- *Recognizing Reportable Issues in a Hospitality Context*: Staff should be encouraged to escalate a variety of concerns, such as an AI-powered chatbot providing guests with consistently incorrect information about hotel services or emergency procedures, a new AI tool for drafting guest service recovery emails generating responses that sound insincere or inappropriate, an internal AI system for analyzing staff performance data flagging individuals based on criteria that seem biased, an AI-generated marketing image that inadvertently offends a cultural group, a scheduling AI that consistently assigns unpopular tasks unfairly, or any instance where AI output seems to violate data privacy.
- *The Benefits: Swift Identification, Human Correction, and Organizational Learning*: Well-defined escalation paths are a mechanism for continuous improvement and risk mitigation. Problems are flagged quickly by those on the front lines, allowing for rapid intervention before a minor glitch becomes a major incident. Once an issue is escalated, it triggers a process where human experts can investigate, validate the concern, and, if necessary, override the AI, correct its output, or temporarily suspend its use, ensuring human judgment remains the final arbiter. Each reported incident is a learning opportunity, providing valuable data points that can inform refinements to AI models, update AI user training programs, fine-tune AI governance policies, and highlight areas where specific AI tools may require more intensive human oversight. This creates a cycle where the organization becomes increasingly adept at deploying AI responsibly.

Investing in your people as AI champions and creating robust, trusted escalation paths ensures that the human element remains central to your AI strategy. This approach significantly changes your team from passive technology users into active participants in its responsible and ethical stewardship, safeguarding your brand and enhancing the human experience that defines hospitality.

Leading with Unwavering Integrity

By proactively defining your hotel's ethical stance on AI, mitigating risks with human oversight, and establishing clear governance, your leadership team builds a resilient foundation for AI success. You foster an environment where teams innovate responsibly, strengthen trust with guests and employees, and ensure technology always amplifies human talent. This commitment to responsible AI, with humans always guiding, verifying, and taking ultimate accountability, is good governance and good business.

This chapter emphasizes that responsible AI in hospitality hinges on strong ethics, responsibility, and safety, all operationalized through robust governance with the human-in-the-loop (HITL) as a central principle. We explored key ethical risks, like bias, hallucinations, IP issues, training data concerns, privacy, sustainability, job displacement, transparency, and accountability, and highlighted proactive, HITL-based mitigation strategies for each. The comprehensive framework for AI governance includes empowering AI champions; establishing clear escalation paths; forming a strategic AI governance committee focused on strategic oversight, responsible innovation enablement, and responsive issue management; and crafting a practical, living AI governance policy. The core message is that thoughtful governance with continuous human oversight enables the adoption of confident, ethical, and impactful AI.

Key Takeaways:

- Effective AI adoption in hospitality demands a strong foundation of ethics, responsibility, and safety, operationalized through a comprehensive AI governance framework with the human-in-the-loop (HITL) as a non-negotiable central design principle.
- Key ethical risks that require proactive, HITL-based mitigation strategies include algorithmic bias, AI hallucinations, intellectual property concerns, training data integrity, data privacy and security, sustainability impacts, potential job displacement, and ensuring transparency and accountability in AI-driven decisions.
- A robust AI governance structure relies on three pillars: an empowered AI governance committee for strategic oversight; a clear, practical, and living AI governance policy addressing tool usage, data handling, and ethical standards (including rules against cheating or misuse); and clear escalation paths to ensure responsible AI use at all levels.
- Thoughtful and continuous human oversight, embedded within the governance framework, enables confident innovation, turning AI into a strategic asset that enhances the guest experience and employee well-being rather than a source of unmanaged risk.

Bringing This into Focus:

- **Initiate an Ethical AI and HITL Discussion with Your Leadership**: This week, review the definitions of responsible, ethical, and safe AI with your core leadership team. Discuss specific scenarios within your hotel where a human in the loop would be critical for decisions, such as handling a VIP complaint, approving a new marketing campaign based on AI insights, or reviewing AI-suggested pricing changes.

- **Conduct a Focused Risk Assessment with HITL in Mind for One AI Tool**: Choose one AI tool or AI-assisted process currently used or considered for your hotel. Using the ethical risk categories from this chapter, identify the top two risks for that specific application. For each risk, brainstorm how current or planned processes effectively mitigate it or where HITL interventions need to be strengthened.
- **Draft Core AI Data Handling and HITL Protocols as a Starting Point**: If your hotel lacks a specific AI policy, begin by drafting a simple, one-page guideline. This initial document should focus on non-negotiable rules for handling guest PII and confidential hotel data with any AI tool and clear expectations that a human must always be in the loop to review and approve any AI-generated content or decision before it impacts a guest or becomes official hotel communication or action. This serves as a foundational step toward a full AI governance policy.

Leadership Action Agenda:

- **Mandate and Resource Your AI Governance Committee and Policy Development**: If not already established, formally charter your cross-functional AI governance committee this month, ensuring it has clear executive sponsorship. Task them immediately with drafting or comprehensively reviewing and updating your hotel's AI governance policy with a specific directive that the human-in-the-loop principle must be explicitly integrated into all relevant sections. Set a clear deadline for the first draft.
- **Design a Human-in-the-Loop Process for a Key AI Use Case**: Select one high-impact AI use case your hotel is actively pursuing. Conduct a dedicated workshop with the relevant team and an AI governance committee member to carefully map out all necessary human-in-the-loop checkpoints for that process. Clearly define who, by role, is

responsible for the review, what criteria they will use, and how overrides or escalations are to be handled. Document this HITL workflow for future reference and training.

- **Develop and Roll Out AI Ethics and HITL Awareness Training:** Create and schedule a concise, mandatory training module for all staff members interacting with or being impacted by AI systems. This training should cover your hotel's core AI ethical principles, key risks like bias and data privacy, the content of your AI governance policy, and their roles and responsibilities as humans in the loop in various scenarios. Ensure all participants understand the why behind HITL to foster a culture of responsible AI use.

9
CRITICAL THINKING IN THE AGE OF AI

Sharpening Your Human Edge

OUR JOURNEY SO far has explored AI's role, the need for AI literacy and mindset, how to identify use cases, understanding the AI toolkit, master prompting, and establishing strong ethical governance. These elements create the framework for using AI effectively. Yet, even with the best tools and guidelines, one human capability remains essential for success: critical thinking. Guiding our AI ascension with wisdom is imperative. This process starts even before you craft a prompt by clearly defining the problem AI is meant to help solve, ensuring it is the right challenge for AI, and considering the quality of data it might access or be trained on.

We now operate in a world where AI can generate information, draft communications, analyze data, and suggest solutions with impressive speed and fluency. It might be tempting to think deep, discerning thought is becoming less important, a task we can increasingly outsource to our digital partners. This temptation leads to a very real phenomenon known as cognitive offloading. Cognitive offloading occurs when we over-rely on external tools, like AI, to do our thinking for us, potentially diminishing our critical thinking abilities, problem-solving skills, or even memory retention over time. Because AI can provide answers and solutions so quickly and plausibly, there is a subtle risk of reducing our mental engagement and accepting AI-generated output without sufficient scrutiny. Although AI is an incredibly powerful tool for reducing toil and augmenting our capabilities, its purpose is to be an intelligent assistant or collaborator, not a replacement for your essential human judgment and deep understanding. The goal is always augmented intelligence, where AI sharpens our thinking. Recognizing this risk of cognitive offloading upfront is vital, as it underscores why actively sharpening your human edge, as this chapter explores, is more urgent than ever.

The more intelligent our AI tools become, the more pressing our human thinking, judgment, and understanding are. AI systems don't understand information, guest situations, or your brand in the way your team does; they are advanced pattern synthesizers working from statistical probabilities in their training data. They lack human-like factual recall, lived service experience, true comprehension of meaning, or an ethical compass aligned with your hotel's values. While AI can be articulate and helpful for accelerating tasks or generating ideas, it is fallible, has incomplete knowledge, and is not ultimately responsible for its suggestions' outcomes. That responsibility always lies with you, your team, and your hotel brand

This means the most vital skill in human-AI collaboration is, above all, consistently exercising sound judgment. It involves building, in yourself and your team, the unwavering habit of pausing before accepting any AI output. It is developing the intellectual discipline to question its assumptions, the analytical courage to challenge its conclusions, especially factual ones, and the professional confidence to correct its errors or override its suggestions based on your deeper operational understanding or ethical considerations. Without this constant human critical review, even the most advanced AI can become an operational liability or a brand risk, not because the technology is flawed but because its output might have been applied without proper human examination. This chapter focuses on sharpening that critical faculty, ensuring human intelligence always leads and wisely uses AI in hospitality.

Understanding AI's "Thinking"

Generative AI excels at processing patterns in massive datasets, automating repetitive tasks, and generating coherent first drafts of almost anything. These are powerful augmentations capable of efficiencies and new creative potential, saving significant time. Still, it is imperative to constantly remember and reiterate what current AI generally lacks. It has an absence of true human intuition born from lived experience. It has an absence of an inherent ethical ethos for novel moral dilemmas. Its "creativity" is largely sophisticated pattern recombination, lacking original inspiration from consciousness or deep emotional understanding. It lacks genuine emotional intelligence to truly empathize or read subtle social cues. And it often misses a common-sense understanding of the physical world or complex human motivations.

For example, ask AI to compose a history of your historic hotel. It might produce a beautifully written, convincing-sounding narrative with plausible names and dates. However, the AI does not recall these from a verified archive specific to your property unless you provide it. It is skillfully assembling common patterns of historical language. It is not lying if details are fabricated; it is performing its function as a pattern-matcher. As a critically thinking hotelier, your indispensable role is to take that AI draft as an interesting starting point and then carefully verify every purported fact against your hotel's actual records before using it. This understanding that AI primarily produces plausible outputs according to patterns is fundamental to engaging with it. Research has even shown that over-reliance on AI's plausible outputs can subtly reduce a user's critical engagement. Our goal is to use AI's power without diminishing our teams' irreplaceable capacity for deep, indispensable thought.

Critical Thinking Beyond Basic Evaluation

Critical thinking with AI involves a more comprehensive content review, which includes:

- **Uncovering Assumptions**: Every AI output is based on assumptions, some from its training data and some from how your prompt was interpreted. A critical thinker asks: What are the unstated assumptions here? Are they valid in our specific hotel context? For instance, if AI suggests a marketing campaign targeting luxury travelers, what assumption is it making about what luxury means to your specific desired guest segment versus a generic definition? Similarly, if AI analyzes staff feedback and suggests a new training program for communication skills, what assumption is it making about the root cause? Could it be unclear processes or a lack of empowerment rather than a skill deficit? Recognizing and questioning these hidden assumptions is essential before acting on AI advice.
- **Evaluating Evidence and Sources when AI Provides Them**: If your AI tool, especially one with web-browse capabilities, cites sources or provides data, your critical thinking must extend to assessing the credibility, timeliness, and potential bias of those sources. Is the cited website a reputable industry journal or a less reliable blog? Is the data current or potentially outdated? Avoid just accepting that AI found it online; critically evaluate where it was found.
- **Recognizing Logical Fallacies or Weak Arguments**: AI can generate text that sounds persuasive. A critical thinker looks deeper: Is the AI's argument truly sound? Does its conclusion follow from the premise? Are there any common logical fallacies, such as generalization from insufficient

data or confusing correlation with causation? Your ability to spot these weaknesses is vital.

- **Considering Alternative Perspectives and Interpretations:** AI often converges on the most statistically probable answer. A critical thinker asks: What other valid perspectives has the AI missed? What alternative interpretations could exist? How might its output change if prompted with a different assumption?
- **Deeply Understanding Contextual Detail Beyond Surface Relevance**: This means asking: How does this specific AI output align with *our* hotel's unique brand identity, service culture, target guest demographics, current market position, strategic priorities, and operational environment? A generic good idea for one hotel might be wrong for yours.
- **Foreseeing Potential Consequences (Second and Third-Order Thinking)**: Critical thinkers do more than just evaluate the immediate output; they think ahead about downstream consequences. If AI suggests a new pricing strategy, what are the potential short-term gains and long-term impacts on brand perception or loyalty? This second-order thinking helps make more robust decisions.

The TUR Filter: A Practical Starting Point for Daily AI Output Evaluation

The TUR filter—evaluating for truth/accuracy, usefulness, and contextual relevance—provides a strong foundation for the rapid assessment of AI outputs.

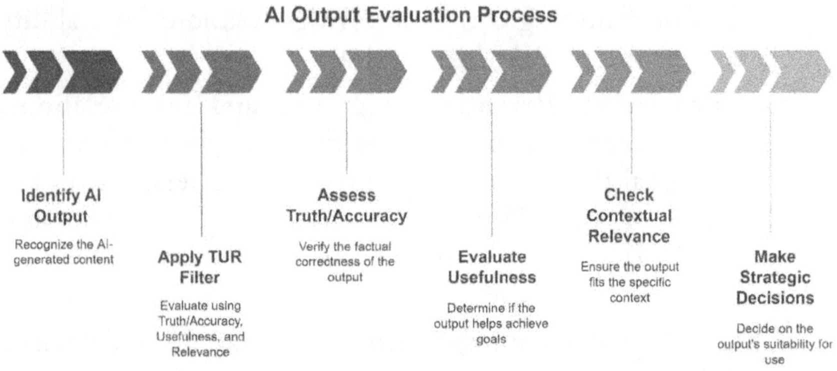

AI Output Evaluation Process

Identify AI Output
Recognize the AI-generated content

Apply TUR Filter
Evaluate using Truth/Accuracy, Usefulness, and Relevance

Assess Truth/Accuracy
Verify the factual correctness of the output

Evaluate Usefulness
Determine if the output helps achieve goals

Check Contextual Relevance
Ensure the output fits the specific context

Make Strategic Decisions
Decide on the output's suitability for use

- **Scrutinize for Truth/Accuracy**: Is this AI output demonstrably true and verifiable? This is non-negotiable for factual statements, operational figures, policies, or safety guidelines. For example, AI drafts a guest email stating your pool is open "24/7." Your manual check against actual hours, which are 7:00 a.m. to 9:00 p.m., prevents guest disappointment. The rule is to always cross-reference AI factual claims with official hotel sources. Remember, AI can hallucinate, confidently inventing information.
- **Evaluate for Usefulness**: Does this AI output help achieve your specific practical goal? For creative or subjective tasks, the question is, "Is this useful?" Does it clarify your thinking and provide a solid foundation, or is it well-written noise? For example, AI generates five generic slogans for your Digital Detox Retreat. Your critical filter asks, "Do these capture our unique essence for stressed professionals? Will they drive bookings?" Discard outputs that lack strategic usefulness.
- **Assess for Contextual Relevance**: Does this AI output truly fit your specific hotel's unique brand, service culture, guest demographics, strategic priorities, and the subtleties of this particular situation? AI is a generalist; your hotel is specific. For example, AI suggests an Employee of the Month program and pizza parties to boost morale. Your

thought filter considers: *Is a pizza party inclusive for our night shift or staff with dietary needs? Is a single award the most effective recognition in our culture?* This ensures AI suggestions are adapted to fit your specific hotel.

Consistently applying these deeper critical thinking principles and the TUR filter empowers you to maintain quality control, ensuring AI serves your objectives and upholds your brand's standards.

Probing AI with Strategic Skepticism

Critical thinking with AI is an active, intellectually vigorous interaction. You are always the human in charge.

When directly prompting an AI assistant, use counter prompting. Imagine AI drafts a proposal for a new hotel sustainability initiative. Avoid just accepting it. Engage in counter prompting, a practice of rigorously challenging plans or outputs by adopting an adversarial or deeply skeptical approach to uncover overlooked flaws or vulnerabilities. Actively look for flaws, unstated assumptions, or unintended negative consequences. Ask, "What aspects of the luxury guest experience might this plan negatively impact if we lack care and that the AI has failed to address? What are the unstated assumptions about retrofitting costs?" This pressure tests the AI's plan by asking it to argue against its own suggestion or to critique an idea you proposed. If AI outlines the benefits of a new flexible cancellation policy, counter prompt: "Now, adopt the persona of our cautious finance director. Articulate all reasons why this policy might fail or have the biggest financial risks." This forces the AI and your team to consider competing perspectives and uncover weaknesses.

Implement a consistent system of regular, strategic spot-checks. If your AI tags online reviews, periodically have managers review batches. Did AI accurately categorize a complaint or interpret sarcasm? These spot-checks are your ongoing validation loop. This active engagement ensures you are validating its outputs.

Prioritizing Relentless Curiosity

A subtle and significant risk with increasingly fluent AI is plausible-sounding AI answers that quietly go unquestioned. When useful information appears instantly, the human tendency is to accept it with relief. This cognitive off-ramp is tempting precisely because AI can reduce our immediate mental workload. Therefore, actively cultivating relentless curiosity requires a conscious effort to re-engage our analytical faculties. This is an erosion of intellectual rigor because it is so easy to use and often "good enough." Complacency occurs when we unconsciously assume AI knows best; we mistake fluency for understanding or rapid processing for wisdom. In a hotel, this culture of AI curiosity means:

- After AI summarizes guest feedback, a curious guest relations manager asks, "What subtleties did this AI summary fail to highlight? Why might those points still be important?"
- When an AI recommends a marketing tagline, a curious marketing director probes, "What assumptions did AI make about our target guest? How can we test those?"
- If an AI-drafted SOP seems logical, a curious training manager discusses with staff, "How would a new employee interpret this under real-world pressure?"
- If an AI suggests service recovery actions, a curious guest service team lead reflects, "Why this combination? Is it optimal for this guest, considering their loyalty or emotional context?"

Fostering this culture of active operational curiosity—always asking, "Why?" "What if?" "What else?" and "How could this be better?" about AI's contributions—is vital. It empowers your team to look beyond AI outputs, question assumptions, collaborate toward superior solutions, and maintain human ownership of the core thinking process. This significantly changes AI from a source of passively accepted answers into a catalyst for deeper human insight and innovation.

Facilitating Critical Thinking Exercises

The following exercises are designed to strengthen your team's critical thinking skills in the context of AI. Consider these facilitation tips to get the most out of these collaborative sessions.

- Create a safe and open environment where team members feel comfortable sharing ideas, asking questions, even seemingly basic ones, and challenging assumptions without fear of judgment. Emphasize that diverse perspectives are valuable.
- Actively solicit input from everyone, including those who might be quieter or offer contrarian views; different roles and experiences will bring unique insights to the evaluation of AI outputs.
- The primary goal is to practice and develop the process of thinking critically about AI. Encourage discussion around how conclusions are reached, what assumptions are being made, and what evidence is considered, rather than rushing to a single "correct" solution.
- Promote collaborative problem-solving by framing the exercises as team challenges and encouraging participants to build on each other's ideas.
- Relate exercises to real-world hotel scenarios whenever possible to make the critical thinking process more relevant and the learnings more applicable.
- If critiquing AI outputs, guide the team to be specific, focusing on how the output could be improved or what its limitations are.
- Keep the exercises focused on maintaining engagement; exploring one or two scenarios is often better than rushing

through many. By fostering this kind of thoughtful and interactive environment, you will help your team internalize the critical thinking habits essential for thriving in an AI-augmented hospitality situation.

Practical Exercises in Critical AI Thinking

These exercises are most effective when conducted collaboratively within teams. This approach fosters discussion, allows for diverse perspectives to surface, and cultivates a shared culture of critical engagement with AI in a psychologically safe environment where questioning and thoughtful analysis are encouraged. Here are practical exercises for your hotel team.

- The Hotel Polish Exercise (Practicing "Trust and Always Tweak"):
    - *Concept*: Move beyond accepting a good enough AI draft to co-creating an exceptionally high-quality, brand-aligned, guest-centric final product.
    - *Practice*: Ask AI to draft a common guest communication, for example, "Write an upsell email for a room booking at [Your Hotel] for a guest who booked a Family Fun Package." Then, challenge your team to collaboratively identify at least five substantive ways to improve that AI draft beyond basic grammar. Focus on: Does it perfectly reflect our specific hotel's unique selling points and policies? Is the tone spot-on? Is vital information missing? Could the call to action be clearer?
    - *Strategic Goal*: To significantly change a competent generic draft into exceptional, personalized communication, saving time on extensive rewrites by making initial refinement more strategic.

- Playing Devil's Advocate with Your AI (The Strategic Pre-Mortem Simulation):

    - *Concept*: Use AI as an intellectual sparring partner to proactively uncover weaknesses or risks in your plans before launch.
    - *Practice*: Use AI to outline a new hotel initiative, for instance, "Outline a plan for a Sustainable Stay and Local Impact program." Once AI generates an initial plan, engage in counter prompting: "That is a positive plan. Now, adopt the persona of a skeptical, financially conservative hotel owner. Or be a convenience-focused leisure guest. From those perspectives, argue against this program. What are its biggest downsides, hidden costs, or reasons guests might have an absence of participation or react negatively?" Observe how the AI articulates counterarguments.
    - *Strategic Goal*: This helps your team use AI as a digital devil's advocate, surfacing potential weaknesses and allowing you to refine your approach for greater success.

- The AI-Assisted Service Audit (Carefully Spotting the Distinction Flaw):

    - *Concept*: Hone your team's ability to critically evaluate AI-generated responses in detailed, emotionally charged guest service and reputation management contexts.
    - *Practice*: Generate five different AI-drafted responses to five genuine, anonymized online guest reviews your hotel received: a glowing five-star, a mixed three-star, one on a specific operational issue, one on price/value, and one seemingly unreasonable. Have

your team analyze each AI response: Does it feel genuinely empathetic and personalized or generic? Does it adequately address all specific concerns? Does any proposed service recovery align with the hotel policy? Are there subtle factual inaccuracies or a tone mismatch?
  - *Strategic Goal*: This builds editorial awareness, spotting when a polished AI response misses human subtlety or fails to convey genuine empathy, ensuring the HITL adds true value to guest interactions.

- The "What Is Critically Missing?" Challenge (The Local Expertise and Insider Knowledge Test):
  - *Concept*: Highlight the irreplaceable importance of your human team's deep local knowledge and contextual experience in enriching AI's often more generic information.
  - *Practice*: Ask a general-purpose AI a question relying on detailed local knowledge, for example, "Generate a list of five unique, family-friendly, budget-conscious activities for a family with young children staying next month at [Your Hotel] in [Your Specific Town/Region]." Review the AI's list with your most experienced concierge staff. Ask, "What unique local gems, off-the-beaten-path experiences, or essential insider tips did the AI miss or get slightly wrong?" For instance, did it overlook the local petting zoo? Did it recommend a museum whose children's wing is closed? Did it suggest a hike without mentioning limited trailhead parking, a key detail that saves guests time and frustration?
  - *Strategic Goal*: Reinforce that although AI provides a good starting point, your team's deep local expertise

significantly changes a generic recommendation into a memorable, perfectly customized guest experience.

These exercises, practiced regularly, develop the mental agility to collaborate with AI effectively, strategically, and responsibly, ensuring your human judgment always guides this powerful technology.

Human Judgment as Your Greatest Asset

This chapter emphasizes that critical thinking is indispensable when working with AI. You must evaluate AI outputs for truth/accuracy, usefulness, and contextual relevance (TUR framework) and apply deeper critical inquiry, which includes identifying assumptions, evaluating evidence, spotting fallacies, considering alternatives, understanding context, and foreseeing consequences. Remember, AI is a pattern-matcher; humans are the truth engine. Utilize counter prompting, plus vigilant spot-checks, and prioritize curiosity over complacency. Investing time in critical thinking upfront saves enormous time and prevents problems, paving the way for a successful AI ascension.

Key Takeaways:

- In the age of AI, human critical thinking is essential; it is your primary tool for navigating AI's capabilities and limitations, including the risk of cognitive offloading.
- The TUR filter is a practical starting point for daily AI output assessment: evaluating for truth/accuracy, usefulness for your specific goal, and contextual relevance to your unique hotel environment.
- Deeper critical inquiry involves uncovering assumptions, evaluating evidence, recognizing logical fallacies, considering alternative perspectives, understanding contextual subtlety, and foreseeing the potential consequences of AI-generated suggestions.

- Actively engage with AI using counter prompting for direct interactions and maintain vigilant spot-checks for automated AI systems, fostering a culture of curiosity over complacency.

Bringing This into Focus:

- **Evaluate an AI Output with Deeper Critical Thinking This Week**: Pick one significant piece of AI-generated content. Assess it using the TUR framework with a colleague and try to identify its underlying assumptions. Consider alternative interpretations the AI might have missed. Discuss what new insights emerge from this deeper analysis.
- **Try an Advanced Critical Thinking Exercise**: Choose one of the practical exercises detailed in this chapter, such as playing devil's advocate with an AI on a new hotel initiative or conducting a "What Is Missing?" local knowledge test with your concierge team. Dedicate thirty minutes to the exercise. Analyze what hidden risks, gaps in AI knowledge, or opportunities for human value-added knowledge this reveals.
- **Practice Probing Questions Systematically**: For any task this week that involves AI assistance, make it a point to ask the AI or yourself about the AI's output. Ask at least two critical follow-up questions that probe its reasoning or assumptions. Examples include, "What data sources most influenced this output?" or "What are the top three limitations of this recommendation?"

Leadership Action Agenda:

- **Develop an Advanced Critical AI Evaluation Training Module**: Enhance your existing AI literacy training (Chapter 2) with a dedicated module focused on advanced critical thinking with AI. This module should cover identifying AI assumptions, evaluating evidence provided by AI, recognizing potential logical flaws in AI output, and considering alternative interpretations. As a practical exercise, include an interactive group evaluation of a complex AI-generated strategic recommendation.

- **Formalize a Counter Prompting AI Initiative Process**: For one significant new AI initiative or a major AI-generated proposal, formally assign a diverse team from different departments. Their task is to rigorously find flaws, potential biases, or unintended negative consequences in the AI's output or the proposed plan. Ensure their feedback is documented and formally incorporated into the project plan. Consider how this team process can become a standard part of your AI project lifecycle to save time and avoid costly mistakes in the future.
- **Champion a Deep Curiosity Culture Regarding AI**: At your next leadership meeting, initiate a discussion on how to cultivate a culture that moves beyond surface-level curiosity about AI to one of deep curiosity. Commit to personally modeling this behavior by consistently asking probing questions that challenge AI's underlying reasoning or data sources. Encourage department heads to recognize and reward team members who thoughtfully question AI outputs and offer superior, human-refined solutions.

PART 4

THE HUMAN ELEMENT IN AN AI-AUGMENTED FUTURE

10
ACHIEVING HUMAN-AI EQUILIBRIUM

AS WE STAND on the edge of disruption with AI integration into every hotel, a critical question, especially for an industry built on human connection, rightly emerges: How do we ensure this AI-powered future genuinely enhances, rather than diminishes, the authentic human touch that defines true hospitality? How do we make this advanced vision practical and human-centric for all types of hotels, including those with constrained resources?

This chapter directly addresses these concerns. While the hospitality industry is still in the early stages of developing a portfolio of replicable case studies demonstrating a healthy human-AI partnership, the principles, strategies, and deliberate choices for achieving this balance are key. We will explore how the humanity-first, AI-powered ethos can be translated into actionable operational strategies, ensuring that AI amplifies human capabilities, reduces toil, and creates more space for the joy of genuine service and meaningful work.

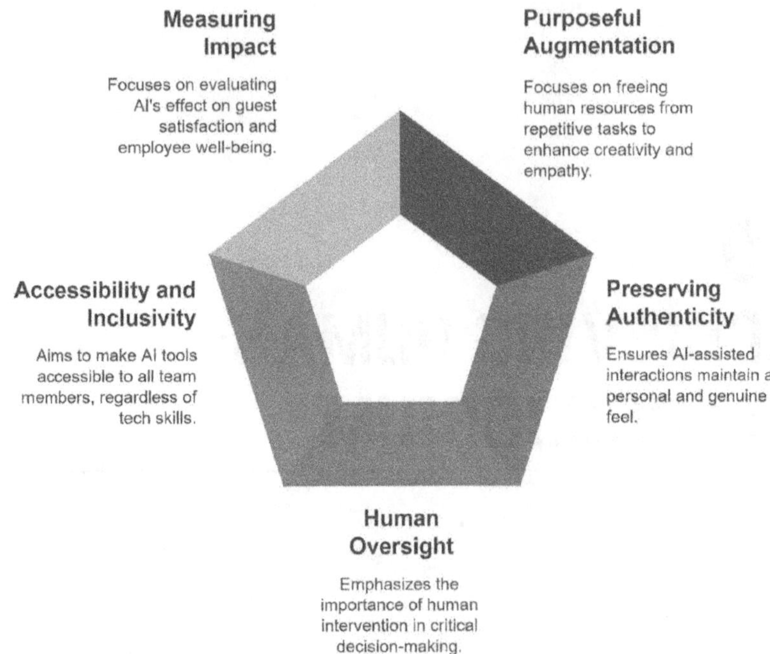

The Humanity-First, AI-Powered Ethos in Daily Practice

The humanity-first, AI-powered principle must be an active filter through which every AI initiative is vetted. Hotel leaders and teams must consciously ask vital questions before and during AI implementation:

- **Purposeful Augmentation**: Will this AI application *genuinely* free up our human team members from specific, identified toil-heavy tasks? How will that newly freed time be reallocated to specific, higher-value, more empathetic, or creative human-centric activities?
- **Preserving Authenticity**: How will this AI impact the perceived warmth, sincerity, and authenticity of our guest

interactions and our brand voice? What measures will ensure AI-assisted communications feel personal and genuine, not generic or automated?

- **Non-Negotiable Human Oversight**: Where are the absolute, non-negotiable points for human-in-the-loop (HITL) intervention, review, and final decision-making?
- **Accessibility and Inclusivity**: How will this AI tool be accessible and beneficial for all team members, regardless of their current tech savvy? How do we avoid creating new digital divides within our workforce? (This links to the skill development discussed in Chapter 11).
- **Measuring True Impact**: Beyond efficiency metrics, how will we measure the impact of this AI on guest satisfaction related to human warmth and staff engagement, employee fulfillment, and reduction of burnout?

By ingraining these questions into the AI evaluation and deployment process, the humanity-first aspect remains at the forefront, ensuring technology serves the core mission of hospitality.

The Process of Balancing in Resource-Constrained Environments

The vision of sophisticated AI agents or extensive custom automation might seem distant for smaller hotels or independent properties. However, achieving a human-AI equilibrium is rooted in a strategic mindset and incremental adoption.

- **Start with High-Toil, Low-Cost AI Wins:** Identify the one or two most time-consuming, repetitive administrative tasks that burden your small team. Could a low-cost AI tool alleviate a significant portion of this? For instance, using AI to draft initial responses to common website inquiries, which are then *always* reviewed, personalized, and sent by a human, can save precious minutes per inquiry, adding up to significant time for a small team.

- **Focus on AI for Internal Efficiency First:** Before considering guest-facing AI, use it to streamline internal processes. Can AI help summarize daily reports, draft staff meeting agendas, or brainstorm ideas for internal staff engagement? Reducing internal toil indirectly benefits guest experience by freeing up staff mental bandwidth.
- **Prioritize Human Training in Critical Thinking with AI**: Even with simple tools, the principles from Chapter 9 (evaluating for truth/accuracy, usefulness, and relevance) are vital. Train your team to be discerning users of any AI, ensuring outputs are always critically reviewed before use.
- **Leverage Shared Prompt Libraries**: As explored in Chapter 5, even a small team can benefit from collaboratively developing and sharing effective prompts for common tasks, ensuring consistency and saving time without initially needing advanced custom GPTs.
- **The One Hour Reinvested Principle:** Encourage team members to identify how one hour saved per day through AI could be reinvested into specific, proactive guest engagement activities or service enhancements. This makes the benefit of AI tangible and directly linked to the core mission.

The balance in resource-constrained environments is about the smart, targeted use of accessible AI to enhance human capacity.

The Joy-Toil Spectrum

Use a human-centric lens to understand AI's potential to reshape work lives: the joy-toil spectrum. Imagine every role in your hotel comprises various tasks plotted along this spectrum.

At one end is toil. Toil encompasses repetitive, mundane, administrative, or data-heavy tasks. These are often the activities that drain people because they typically do not require significant thought, critical problem-solving, or the application of unique human skills like empathy, creativity, or strategic reasoning. Examples include manual data entry, drafting nearly identical emails, compiling basic reports, or routine scheduling. This type of work consumes energy and fails to contribute significantly to an employee's sense of fulfillment or directly enhance guest value.

At the other end lies joy. Joy includes work that brings professional satisfaction, utilizes our highest human capacities, and creates lasting value for the individual, the team, and the guests. These are often the very activities that attract individuals to their specific roles in hospitality and generate a sense of purpose and accomplishment. It manifests in personalized guest interactions, creative problem-solving, strategic decision-making, mentoring colleagues, innovating new guest experiences, and building a genuine community within the hotel and with its clientele.

Measuring Toil and Joy

Individual tasks within a role can be assessed to make this spectrum actionable. Consider using a simple one-to-five scale for both dimensions:

- **Toil Score:** Rate each task from one (*low toil:* engaging or neutral effort, may involve some thought) to five (*high toil:* draining, highly repetitive, requires minimal thought, unfulfilling). A high score indicates a prime candidate for AI augmentation or automation.
- **Joy Score:** Rate each task from one (*low joy:* no satisfaction, feels disconnected from role's core purpose) to five (*high joy:* energizing, utilizes best skills, directly aligns with why one chose the role, feels valuable and impactful).

One of the objectives of AI integration is to significantly shift the balance by reducing tasks with high toil scores, thereby creating capacity for tasks with high joy scores.

Hotels can begin to track this shift systemically. One method involves incorporating questions derived from this one-to-five scaling into employee engagement surveys or specific work-analysis workshops. These questions would assess perceived levels of toil versus joy associated with core responsibilities. Conducting these assessments before and after targeted AI interventions provides measurable data on the impact of AI on the quality of work life.

AI's promise in hospitality is its ability to systematically reduce the toil prevalent in many roles. AI can free up human talent by intelligently automating or augmenting energy-draining tasks that require little thought. This liberation of time, cognitive bandwidth, and emotional energy allows employees to dedicate themselves more fully to activities that generate professional joy. Such activities contribute meaningfully to individual professional growth and overall team morale.

AI as a Facilitator of Connection, Not a Substitute

A core principle in achieving balance is viewing AI as a powerful assistant that handles the impersonal, data-intensive aspects of understanding and preparation, *enabling* humans to connect more effectively and authentically. Imagine AI sifting through guest feedback from multiple sources (surveys, online reviews, social media) to identify recurring themes or emerging issues (an analytical AI function as per Chapter 3). This AI does not feel the guest's frustration but can present a clear, data-backed summary to the guest relations manager. Armed with this AI-generated insight (e.g., "Multiple guests this week have commented on slow breakfast service between 8:00–9:00 a.m."), the manager can then engage with their team and affected guests with a deeper understanding of the problem, leading to more targeted and empathetic human-led solutions. The AI provides the "what;" the human delivers the "why it matters" and the "how we compassionately respond."

Training for Balance

Achieving human-AI equilibrium requires specific training that goes beyond tool proficiency. This involves:

- **Cultivating Critical AI Literacy (Chapter 9):** Continuously reinforcing the skills to question AI outputs, identify potential biases, and understand AI's limitations.
- **Developing AI Collaboration Skills:** Teaching staff how to effectively brief their AI remote associate (Chapter 5), how to iterate with AI to refine outputs, and, importantly, when to *stop* using AI and rely on human judgment.
- **Ethical Decision-Making in AI-Augmented Scenarios (Chapter 8):** Providing frameworks and scenario-based training to help staff navigate ethical gray areas where AI

might offer a solution that is efficient but not optimally aligned with guest well-being or hotel values.

- **Empathy and Emotional Intelligence Reinforcement:** As AI handles more routine tasks, intentionally focus training on enhancing the uniquely human skills of empathy, active listening, and emotional intelligence, the very skills AI cannot replicate.

Championing and Modeling the Human-AI Equilibrium

Ultimately, hotel leadership is responsible for fostering and maintaining a healthy human-AI balance. This requires:

- **Articulating a Clear Vision**: Consistently communicating that AI's purpose is to *empower* humans and enhance the guest experience.
- **Investing in Human Development:** Prioritizing the training and upskilling necessary for teams to thrive alongside AI.
- **Modeling Balanced Behavior:** Leaders should visibly demonstrate how they use AI as a supportive tool while always applying their critical judgment and emphasizing human connection.
- **Creating Psychological Safety**: Fostering an environment where staff feel safe to experiment with AI, report AI errors or ethical concerns without fear of reprisal, and voice opinions about how AI impacts their work and guest interactions.
- **Celebrating Human-Centric Successes:** Highlighting and rewarding instances where AI *enabled* exceptional human service or innovation, reinforcing the desired balance.

An Ongoing Journey Toward Symbiosis

We can navigate this transformation by consciously focusing on strategies that preserve authenticity, empower existing staff (even those in resource-constrained environments), use the joy-toil spectrum as a guide, and champion AI as a facilitator of deeper human connection. This considered approach ensures that as our hotels become more technologically advanced, they also become more profoundly human, proving that the true promise of AI in hospitality lies in its ability to help us deliver exceptional service with cutting-edge intelligence and enduring heart.

This chapter focused on the strategies for achieving a human-AI equilibrium, ensuring that AI adoption enhances, rather than diminishes, authentic hospitality and empowers staff, even in resource-constrained settings.

Key Takeaways:

- Achieving a human-AI balance requires a deliberate commitment to the humanity-first, AI-powered ethos, where every AI initiative is vetted for its impact on genuine human connection and its ability to free up staff for more meaningful work.
- Resource-constrained hotels can achieve balance by focusing on accessible AI tools for high-toil tasks, prioritizing internal efficiencies, training for critical AI thinking, and leveraging shared prompt libraries, proving that mindset and strategy are as important as advanced technology.
- Continuous leadership commitment, ongoing staff training in AI collaboration and ethics, and using the joy-toil spectrum as a metric are essential for fostering a culture where AI genuinely facilitates deeper human connection and operational excellence.

Bringing This into Focus:

- **Review One AI Tool for Its Human Impact:** This week, select one AI tool or process you currently use. With your team, discuss how it affects the human side of your work. Does it help you connect more with guests, or does it create distance? What is its real impact on your daily tasks and job satisfaction?
- **Design a Balanced AI Solution for a Specific Role:** Choose one job role in your department. Brainstorm how a simple AI tool could take over a specific, repetitive task for that role. Then, clearly identify what new, higher-value human activity that person could do with the time saved.
- **Pilot One Small AI Quick Win:** If your resources are limited, identify one simple way to test AI's benefits on a small scale next month. For example, use a free AI tool to create first drafts for a common type of email, which a human then personalizes, or start a shared document with just one or two effective prompts for a recurring task.

Leadership Action Agenda:

- **Incorporate Human Impact Reviews into AI Governance:** Update your formal process for approving any new AI tool or workflow. Mandate that every proposal must include a "Human Impact" section detailing how the initiative will maintain human oversight and specifically how it is expected to free up employees for more meaningful work or enhance direct human service.
- **Create a Feedback Program for AI's Real-World Effects:** Launch an initiative to gather real-world stories from your staff about how AI is affecting their work, both positively and negatively. Use a safe and perhaps anonymous channel to collect this feedback. These insights are invaluable for

refining your AI strategy and showing your team that their experience matters.

- **Champion Training on Balanced Human-AI Collaboration:** Develop and lead training sessions that focus specifically on achieving a healthy balance between AI efficiency and the human touch. Go beyond teaching how to use a tool and focus on developing your team's judgment about *when* to use AI as an assistant and *when* to rely on their unique human skills for empathy, creativity, and guest connection.

11
FUTURE ROLES

Essential Skills and Strategic Talent Transformation

LET'S LOOK OUTSIDE hospitality for a moment to understand how leaders are starting to think about AI in their companies. Top executives are speaking plainly about AI's impact on work. Duolingo's CEO, Luis von Ahn, is shifting to an AI-first approach, replacing contractors with AI for tasks that can be automated. LinkedIn's CEO Ryan Roslansky predicts AI will accelerate workforce learning, emphasizing that by 2030, 70% of jobs will look dramatically different. And LinkedIn co-founder Reid Hoffman sees the decline of traditional nine-to-five work, with AI enabling more project-based, flexible arrangements. The message is clear: Skills trump job titles, adaptation beats resistance, and human-centric abilities remain invaluable despite AI's growing reach. This global shift in understanding work directly informs our vision of the AI Nexus Hotel, as we'll discuss in Chapter 15. The emergence of human-agent teams, the agent boss role for many employees, and the reconfigured work chart all necessitate a profound transformation in talent and skills within hospitality.

So, what does this mean for our industry? As AI weaves itself into hotel operations, reshaping guest experiences, commercial strategies, back-office functions, and brand messaging, a fundamental question emerges: What does this technological shift mean for us, the people who are the heart and soul of hospitality? This chapter is about a future where AI serves as a powerful amplifier of human capability, thoughtfully reshaping roles for the better and demanding new, uniquely human, and strategically vital skills. It addresses how we, as hospitality leaders, can proactively guide this development. Our goal is to develop talent to thrive alongside these intelligent technologies, ensuring AI empowers our teams to deliver even more exceptional and human-centric guest experiences.

The Evolution of Traditional Hospitality Roles

AI will significantly change existing positions, augment capabilities, and shift focus from routine execution to higher-level strategy and data-informed human interaction.

- The general manager develops from day-to-day operational oversight to chief AI visionary, ethical governor, and cultural leader of AI-augmented teams. They use AI-driven insights for strategic decisions, fostering responsible AI innovation.
- The revenue manager transitions from manual data analysis to strategically defining parameters for AI-powered dynamic pricing and demand forecasting engines. They become an expert interpreter of AI-identified market signals, focusing on long-term revenue strategies.
- The director of marketing becomes an architect of AI-driven personalization at scale, coordinating AI-generated content strategies, using AI for guest segmentation, and guarding ethical AI use and brand voice.

- The front office manager leads collaborative human-AI teams, managing complex guest exceptions escalated by AI and coaching staff on delivering AI-augmented, empathetic service.
- The director of housekeeping oversees AI-optimized resource deployment, such as dynamic staffing and linen allocation, and leverages AI for predictive maintenance alerts.
- The food and beverage director uses AI for dynamic menu engineering, AI-driven demand forecasting for inventory optimization to reduce waste, and personalized F&B recommendation engines.
- The human resources director uses AI for smarter talent acquisition, personalized employee learning pathways, and AI-driven analytics for proactive retention strategies, freeing HR for strategic people development.
- The director of finance uses AI's predictive capabilities for sophisticated financial modeling, AI for real-time anomaly detection, and dedicates more expertise to proactive, data-backed strategic counsel.

This AI-driven development elevates these roles, empowering professionals to shed toil for more strategic and fulfilling work.

The Rise of Specialized AI-Focused Roles

The increasing sophistication of AI will lead to new specialized roles bridging technology and hotel operations. Key roles include:

- **AI Hospitality Strategist:** This senior leader works with the executive team to align AI initiatives with overarching business goals such as RevPAR, guest loyalty, efficiency, and brand differentiation. They architect the hotel's AI transformation roadmap.

- **Hotel Process Automation Specialist/AI Workflow Architect:** This role combines technical skills with deep operational understanding to design, build, test, and maintain AI-driven automated workflows using no-code/low-code platforms. The goal is to streamline operations, eliminate friction, and improve data flow.
- **Guest Experience AI Personalization Manager:** This manager uses AI and guest data ethically and with strict privacy safeguards to design systems delivering hyper-personalized interactions, customized offerings, and customized communications across the guest journey.
- **Hospitality Prompt Engineer/AI Content Coordinator:** This individual master's language, logic, and creative communication to craft, refine, test, and manage effective prompts for AI tools, ensuring optimal performance, brand voice alignment, and accuracy.
- **AI Training and Adoption Lead (AI Change Ambassador):** This person champions AI readiness and designs and delivers AI literacy and mindset training. They support staff adoption, nurture AI champions, and effectively spread skills.

Building Your AI-Ready Workforce

Changing team capabilities requires a comprehensive, ongoing strategy, including creating feedback loops from humans and AI systems to help individuals refine AI collaboration skills. Consider these pathways:

- Design tiered, structured internal training programs, for example, Level 1: AI Essentials; Level 2: Applied AI for specific departments; Level 3: Advanced AI Collaboration and Innovation.

- Launch focused, cross-functional AI sprint projects that tackle specific challenges with AI over six to eight weeks to build practical skills and yield innovative solutions.
- Nurture your internal AI champion network to empower peer coaches who share best practices and model responsible AI use.
- Develop accessible micro-learning AI resources, like short video tutorials, quick reference cheat sheets, interactive quizzes, or a regular AI prompt of the week.
- Implement strategic reverse mentoring programs pairing digitally native junior staff with senior leaders. Successful programs often benefit from clear objectives, a structured approach with regular check-ins, and fostering an environment of mutual respect and open learning.
- Forge external partnerships for specialized expertise by collaborating with AI tech providers, specialized trainers, universities, or industry associations like HSMAI or HFTP.

The 5 A Model for AI Skill Development

To structure your AI talent transformation, consider the 5 A Model. This model is designed to guide leadership through the systematic process of assessing current capabilities, architecting future roles, and acquiring the necessary talent and skills. The framework then moves into the practical application of these skills through targeted activation and creates systems to amplify and sustain continuous learning across the organization. By following these five stages, you can create a clear and actionable roadmap for building a workforce that thrives.

1. **Assess**: Understand current AI readiness. Evaluate roles against the joy-toil spectrum. Assess AI literacy and mindset and skill gaps. *Key Question*: Based on our current team's AI literacy and the toil identified, where are our most immediate skill gaps for leveraging AI effectively?

2. **Architect**: Strategically design your desired future-state organization. How will roles develop? What new AI specialist roles are needed? How will team structures adapt to human-AI partnerships? *Key Question*: How will traditional roles like front desk agent or sales manager need to be redesigned to incorporate AI collaboration and focus on higher-value human interactions in our AI Nexus Hotel model (Chapter 15)?
3. **Acquire**: Translate the blueprint into action. Which skills can be built internally through training or mentorship? Which may need external hiring or partnerships? For external hiring, be prepared for a competitive market for talent skilled in both hospitality and AI; job descriptions may need to emphasize aptitude for AI collaboration and continuous learning, with hotels investing in industry-specific AI training post-hire. How will recruitment strategies attract AI-savvy hospitality talent? *Key Question*: What is our strategy for sourcing AI-specific skills, primarily upskilling existing talent, hiring new specialists, or leveraging external consultants/partners?
4. **Activate**: Implement AI learning initiatives, provide hands-on practice in real hotel scenarios, and foster a supportive environment for skill acquisition and application. This includes just-in-time support and positive reinforcement. *Key Question*: How can we create ongoing, practical opportunities for teams to apply new AI skills to real hotel tasks in a supportive environment that encourages experimentation?
5. **Amplify**: Create sustainable, organization-wide systems for continuous AI learning and knowledge sharing. How can you capture and disseminate best practices? How can you reassess developing skill needs? How can you recognize and reward AI innovation and leadership? *Key Question*: What mechanisms will we establish to share AI best practices

and innovative use cases across departments, ensuring continuous organizational learning?

Building Your Future-Ready, AI-Augmented Hospitality Workforce

Proactively addressing role development and committing to continuous skill development are strategic imperatives. Leading hospitality brands will view their AI journey as a human-centered organizational transformation, using lenses like the joy-toil spectrum to enhance work quality and fulfillment. It involves fostering AI literacy and an adaptive AI mindset and creating development pathways that empower every team member. By doing so, you build teams that are more efficient, effective, engaged, and capable of extraordinary innovation.

To further guide your immediate action, consider these leadership principles:

- Lead with a clear vision by articulating why AI is integrated, emphasizing guest and employee experience enhancement.
- Invest continuously in your people by prioritizing the ongoing development of AI literacy, mindset, critical thinking, and ethical discernment.
- Embrace responsible experimentation by cultivating a culture where teams explore new AI applications and learn within ethical guardrails.
- Balance action with human-centric consideration by ensuring every AI deployment is preceded by careful thought about its ethical and human impact.
- Celebrate progress by visibly recognizing achievements in AI skill development and implementation.

The AI-augmented hotelier is an emerging reality: empowered, strategic, and more capable than ever of focusing on the art of exceptional hospitality. Building this future-ready workforce is a rewarding leadership opportunity.

This chapter painted a vision of the AI-augmented hotelier, where AI significantly changes roles by reducing toil and amplifying joy, leading to more strategic human contributions. We explored the development of traditional positions and new specialized AI roles. We detailed skill development pathways and the 5 A Model (assess, architect, acquire, activate, amplify) as your framework for talent transformation.

Key Takeaways:

- Successfully navigating AI talent transformation requires strategic workforce planning, which can be guided by the 5 A Model (assess current readiness, architect future roles, acquire necessary skills, activate learning and application, and amplify continuous improvement and knowledge sharing).
- Continuous skill development is the new reality, involving pathways like tiered internal training, AI sprint projects, nurturing AI champions, providing micro-learning resources, considering reverse mentoring, and forming external partnerships to build a future-ready, AI-augmented hospitality workforce.

Bringing This into Focus:

- **Identify a Targeted Skill Development Need for Your Team:** Considering the core AI skills and developing hospitality roles discussed, choose one specific skill (e.g., advanced prompting, AI ethics awareness, data interpretation) that you believe is most immediately needed by your team or department. Brainstorm one practical activity or resource from the skill development pathways section that you could introduce next quarter to start addressing this need.

- **Initiate a 5 A Model Starting Question for Your Hotel:** Using the assess phase of the 5 A Model as a starting point, formulate one key question about your hotel's current leadership alignment, workforce capabilities, or organizational culture concerning AI adoption. Plan to discuss this question with your leadership team to better understand your hotel's true starting point for AI talent transformation.

Leadership Action Agenda:

- **Pilot the 5 A Model for a Target Department or Essential Skill Set**: Select one key department ripe for AI integration or one essential AI skill set (e.g., AI-assisted data analysis for revenue management or AI-powered content creation for marketing) critical for your hotel's future. Task a cross-functional working group, including HR/L&D, IT, and relevant departmental leaders, by applying the full 5 A Model (assess, architect, acquire, activate, amplify) to this target area over the next six months. Ensure they develop a clear action plan with measurable objectives for talent development.
- **Develop and Launch a Future Skills and Roles Communication Plan:** Create a clear, proactive, and ongoing internal communication plan to discuss with all team members how AI is expected to develop roles, create new skill development opportunities, and potentially enhance career pathways within the hotel. Consistently emphasize AI as a tool for human augmentation, designed to make their work more fulfilling and impactful rather than a threat to their positions. This transparency is key to effectively managing change.

PART 5

ADVANCED AI SYSTEMS AND STRATEGIC VISION

12
CUSTOM GPTS

OUR JOURNEY HAS taken us from understanding AI basics and building the right mindset to selecting tools and mastering prompts. We have explored how AI enhances your everyday software and creative work, all guided by ethical governance and critical human thinking. Now, we have reached a significant development stage of your AI journey: crafting AI assistants specifically customized to your hotel's unique needs. Instead of repeatedly briefing a general AI, imagine deploying a specialized AI teammate already fluent in your brand voice, SOPs, and guest service philosophy, an AI that saves your team significant time and significantly changes operational drudgery into focused, meaningful work.

Leading platforms, like OpenAI with its Custom GPT feature, allow hoteliers lacking coding skills to create these specialized AIs. Google is moving in this direction, too, with concepts like Gems for personalizing Gemini experiences. Although the environment will develop, the principles of creating these tools are becoming fundamental. This chapter provides a practical guide, focusing primarily on building custom GPTs within ChatGPT, as its features are well-established as of summer 2025.

The Value of Customization: AI That "Gets" Your Hotel

Custom AI assistants operate on two powerful principles: deep specialization and accessible embedded knowledge. You create a persistent AI assistant primed with your specific information and operational directives. As the creator, you define several key aspects: your hotel's precise brand voice and tone guidelines; key operational documents, such as SOPs, training checklists, or rate sheets; detailed property-specific information, like amenities, room types, or menus; and preferred formats and structures for common outputs. Once built, your team interacts with an AI that already understands your hotel, delivering more relevant, accurate, and brand-aligned help, saving everyone time. These custom GPTs can be securely shared within enterprise accounts, ensuring consistency and safeguarding information.

A Step-by-Step Guide to Creating Your First Custom GPT

Although interface details can change, the general process for creating a custom GPT within a ChatGPT Plus, Team, or Enterprise account is straightforward. Always follow your hotel's AI governance policy regarding data security and approved tools.

1. **Access the GPT Builder**: Log in to your eligible ChatGPT account. Look for "Explore GPTs" or similar in the sidebar. Click "+ Create" or "Create a GPT" to reach the builder interface.
2. **Understand the Builder Modes**: "Create" vs. "Configure": The GPT builder usually offers two main tabs.

    - *Create*: An interactive mode where you chat with the builder AI. You describe what you want, and it suggests a name and profile picture and drafts initial instructions. This mode is good for beginners because it is a guided process.

- *Configure*: This mode gives you direct control with specific fields to fill in: name, description, instructions, conversation starters, and options to upload knowledge files and enable capabilities.
- *Recommendation*: Start with "Create" for initial ideas, then switch to "Configure" to fine-tune every detail. For robust, reliable custom GPTs, you will want to spend most of your time in "Configure."

3. **Define Your Custom GPT in "Configure" Mode:**

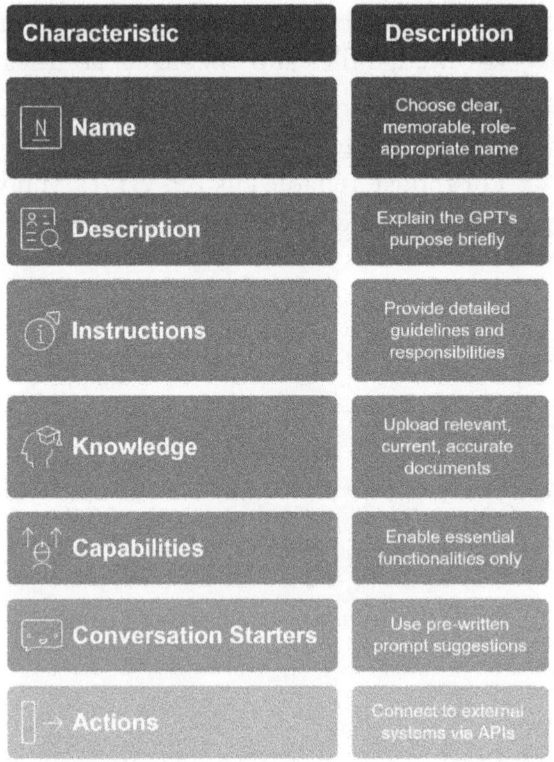

Custom GPT Configuration Options

Characteristic	Description
Name	Choose clear, memorable, role-appropriate name
Description	Explain the GPT's purpose briefly
Instructions	Provide detailed guidelines and responsibilities
Knowledge	Upload relevant, current, accurate documents
Capabilities	Enable essential functionalities only
Conversation Starters	Use pre-written prompt suggestions
Actions	Connect to external systems via APIs

- *Name*: Choose something clear, memorable, and appropriate for the role. For example, "[Your Hotel Name] Concierge AI" or "CRM Eblast Writer" is better than "HotelBot1."
- *Description*: Write a brief explanation of what your Custom GPT does to help users understand its purpose, such as "Assists with drafting on-brand guest communications and answering common questions for The Serene Shores Resort."

- *Instructions (The Prime Directive):* This is the most critical part. Provide detailed, foundational guidelines for your custom GPT. Think of it as the comprehensive job description for your new AI remote associate. Clearly define its role and responsibilities within your hotel, desired personality traits, the specific tone of voice it must adopt, any rules or boundaries that cannot be negotiated, and detailed instructions on how it should respond in common scenarios. Specify information sources it should prioritize from its knowledge base or explicitly state what kind of information it should avoid providing, even if asked.
- *Knowledge (Uploading Files for its "Brain Food"):* Give your custom GPT specialized expertise by uploading relevant documents. Quality matters here: "Good data in, good data out." Ensure uploaded files are current and accurate. For instance, avoid uploading outdated SOPs from three years ago or multiple conflicting versions of your brand voice guide, as this will only confuse your custom GPT and lead to unreliable outputs. Treat uploaded knowledge files with the same data security and privacy diligence as any other hotel document. Ensure no overly sensitive guest PII or confidential strategic information that should lack broad accessibility, even internally, depending on who the GPT is shared with, is included without adherence to your AI governance policy. Examples to upload include your hotel's brand voice guide, key SOPs, excellent email response examples, property information sheets, comprehensive FAQs, and a glossary of hotel-specific jargon.
- *Capabilities (Activating its Superpowers):* Enable or disable specific functionalities. Web browsing allows

the GPT to access live internet information, which is useful for a concierge GPT needing current event information—but use it with caution. DALL·E image generation enables original image creation from prompts. Code interpreter allows the AI to write and execute code for data analysis, charts, or complex problems. Select only capabilities essential to your GPT's defined role.

- *Conversation Starters:* These pre-written prompt suggestions appear when users open your custom GPT. They guide users and showcase the GPT's capabilities.
- *Actions (Connecting to the Outside World):* This more advanced feature elevates your custom GPT's potential. "Actions" allow your GPT to interact with external systems and data sources beyond its uploaded knowledge. Your GPT can "do things" by connecting to other software via APIs. Given that setting up "Actions" often involves API configurations that can be more technical, consider partnering with your IT team, a tech-savvy colleague who enjoys exploring such integrations, or an external AI consultant if you identify a high-value use case where direct system interaction would significantly boost your custom GPT's utility.

4. **Test and Iterate in the "Preview" Pane**: As you configure, a "Preview" pane lets you test your GPT in real time. Ask questions related to its purpose and uploaded knowledge. According to performance, refine its instructions, upload more knowledge, or adjust the conversation starters. This iterative, thorough testing-and-refinement process informed by diverse test scenarios is crucial for building an effective and reliable custom GPT.

5. **Publishing and Sharing Your Custom GPT**: Save and publish with appropriate access once satisfied. Options include "Only me," where only you can access it; "Anyone with a link," where only people with the specific URL can access it; or "Public," making it discoverable by anyone in the GPT Store if enabled. For most hotel custom GPTs with proprietary information, choose "Only me" or "Anyone with a link." Always select the sharing option that aligns with your hotel's data security policies.

Remember that custom GPTs are not "set it and forget it." Plan to periodically review and update their instructions and knowledge files to reflect any changes in your hotel's procedures, branding, or offerings to ensure they remain accurate and effective. Think of your custom GPTs as dynamic digital team members; just like any employee, their knowledge and directives need to stay current.

Consider these key triggers for initiating a review and update cycle for your custom GPTs:

- **Scheduled Periodic Reviews**: Establish a regular cadence for proactively checking your custom GPTs. For many hotel applications, a quarterly review is a good starting point. For more frequently used custom GPTs, a monthly check-in might be appropriate.
- **Major Hotel SOP or Policy Changes**: Whenever there are significant updates to your hotel's standard operating procedures (e.g., new guest check-in protocols, emergency response plans, service recovery guidelines) or changes to key guest-facing policies (e.g., cancellation, privacy), any custom GPT that references or relies on this information must be updated promptly.
- **Branding, Marketing, or Product Offering Updates**: If your hotel undergoes a rebranding, launches new marketing campaigns with distinct messaging, introduces new services and amenities, or significantly alters existing ones like F&B

menus, your custom GPTs, especially those involved in guest communication or content generation like a "Brand Guardian," need to be refreshed with the latest information and stylistic guidance.

- **Observed Performance Degradation or Inaccuracies**: If team members report that a custom GPT is starting to provide consistently off-target, outdated, or factually incorrect advice or if its responses are no longer aligned with current practices, it is an immediate signal that its instructions or uploaded knowledge files require thorough review and correction. Encourage users to provide feedback on the custom GPT's performance.
- **Changes in Integrated Systems (Relevant for "Actions")**: If your custom GPT uses "Actions," any significant API changes or structural modifications to connected external systems may necessitate a review and potential adjustment of the custom GPT's action configurations to ensure continued functionality.
- **Updates to the Underlying AI Platform/Model**: When the core AI platform, for example, OpenAI's models, undergoes major updates or releases new features relevant to custom GPTs, it can be beneficial to revisit your creations to see if they can leverage new capabilities or if their behavior needs slight adjustments to remain optimal. Regularly maintaining your custom GPTs ensures they remain trusted, effective, and reliable tools that save time and enhance the consistency and quality of your hotel's operations and communications.

Practical Example: Creating a "[Your Hotel Name] Brand Guardian" Custom GPT

Let's walk through creating a custom GPT designed to help your team ensure all written communications strictly adhere to your hotel's unique brand tone, style, and voice. This directly helps manage brand consistency and saves significant time in editing and reviews.

1. **Access the GPT Builder**: Follow Step 1 from the previous section.
2. **Configure Tab:**

    - *Name*: "[Your Hotel Name] Brand Guardian" (e.g., "The Grand Plaza Brand Guardian").
    - *Description*: "Your expert assistant for ensuring all text aligns with [Your Hotel Name]'s official brand voice, style, and tone guidelines."
    - *Instructions*: "You are the official brand guardian for [Your Hotel Name]. Your primary purpose is to review any text users provide and ensure it strictly adheres to our brand voice, style, and tone guidelines as detailed in the uploaded knowledge files. Our brand voice is [describe your core brand voice attributes, e.g., sophisticated but warm or elegant and approachable]. Our tone should generally be [describe general tone, e.g., welcoming, respectful, and guest-centric]. Still, adapt tone appropriately for specific contexts if indicated in the user's prompt or implied by the text type. When a user provides text for review, analyze it against the principles in our uploaded 'Brand Voice & Style Guide.' Provide constructive feedback on areas that deviate from our brand standards. Suggest concrete rephrasing or edits to improve alignment. Briefly explain why the suggested changes are recommended, referring to principles from the brand guide if possible. If the text is already well-aligned, affirm this positively. Your overall interaction style with the user should be helpful, precise, encouraging, and professional. Avoid generating new content unless specifically asked

to rephrase or rewrite a provided text according to brand guidelines. Your main role is reviewing and guiding. Refrain from discussing topics outside of brand voice, style, and tone unless directly related to improving a piece of communication for [Your Hotel Name]."

- *Knowledge (Upload Files)*: Your hotel's official, current "Brand Voice & Style Guide" (PDF or Word); five to ten examples of excellent, on-brand marketing copy; examples of off-brand or undesirable phrasing/tone to avoid; optionally, a concise list of key brand adjectives and positive brand phrases.
- *Capabilities*: Web Browse: OFF (as it lacks necessity for this function); DALL·E Image Generation: OFF; Code Interpreter: OFF.
- *Conversation Starters*: Examples include "Can you review this guest email for brand voice alignment?" "Is this social media post aligned with [Your Hotel Name]'s tone?" "Help me rephrase this sentence to sound more [e.g., luxurious and welcoming]." "Does this paragraph from our website meet our style guidelines?"

3. *Test and Iterate*: Use the "Preview" pane extensively. Paste in various pieces of text, some on-brand and some deliberately off-brand, and see how the brand guardian responds. Refine the instructions until they consistently provide accurate and helpful guidance.
4. *Publish:* Once satisfied, publish it, likely as "Anyone with a link" if using a Team/Enterprise account, so your marketing, sales, and guest relations teams can easily access it.

Now, any team member can quickly paste their draft communications into this "Brand Guardian" GPT and get instant, expert feedback on brand alignment. This saves individual writers and their managers time while ensuring brand consistency across all communications.

It is also important to acknowledge current limitations. Custom GPTs, especially those without "Actions" connecting them to live data, rely on the knowledge you provide during creation. This knowledge does not update automatically. Therefore, your custom GPT might need regular file updates or be designed to direct users to live sources for information that changes frequently, like daily rates or specific event schedules. As always, human oversight of their output remains essential.

From Specialized Assistant to Proactive Operator: A Glimpse into Automation

You have now crafted your specialized AI teammates. They understand your hotel, speak your brand voice, and can assist your human teams when prompted. This is a huge step. What is next? For many hotels, it is intelligent workflow automation. We will explore this in detail in Chapter 13, showing how LLMs can be connected to other hotel systems to automatically trigger tasks driven by real-world events. For now, mastering the creation of these custom GPTs provides the essential intelligent "brains" for that future automation, marking a significant milestone in your AI journey.

Key Takeaways:

- Custom GPTs offer deep specialization by utilizing embedded knowledge, such as your hotel's brand guides, SOPs, and property information, to save team time and ensure consistent, on-brand AI assistance, moving beyond generic AI responses.
- The process of building a custom GPT, particularly in platforms like ChatGPT, generally involves accessing a builder interface, defining its name and description, writing detailed

"Instructions" as its prime directive, uploading "Knowledge" files securely, selecting appropriate "Capabilities," setting up "Conversation Starters," and optionally configuring "Actions" for external system integration.

- Thorough testing and iteration using a preview function are essential for refining a custom GPT's performance. Equally important is establishing a plan for periodic updates to its knowledge and instructions to maintain accuracy and effectiveness over time as information and hotel procedures change.
- While custom GPTs provide significant value, their knowledge is static; therefore, human oversight of outputs and understanding their limitations are imperative for responsible use.

Bringing This into Focus:

- **Identify One Potential Custom GPT Use Case for Your Department**: Think about one recurring task or frequent information need within your department where a specialized AI assistant, pre-loaded with specific hotel knowledge, could save significant time or improve consistency. What would its primary purpose be?
- **Outline Core Instructions for Your Custom GPT Idea**: For your identified custom GPT concept, draft three to five core bullet points for its "Instructions" section. Consider: What is its specific role? What primary functions should it perform? Are there topics or actions it should explicitly avoid?
- **List Essential Knowledge Files for a Custom GPT:** Identify two to three key documents (e.g., SOPs, brand guides, templates), ensuring they are current and accurate, that would be essential for this custom GPT's "Knowledge" base.

Leadership Action Agenda:

- **Sponsor a First Custom GPT Creation Workshop for a Pilot Department**: If your hotel utilizes a platform supporting custom AI creation, select one department to pilot developing their first custom GPT for a well-defined, high-value internal use case (e.g., "SOP Helper," "Internal Event FAQ Bot"), emphasizing data security for uploaded knowledge.
- **Develop a Hotel-Wide Custom AI Use Case Registry and Sharing Protocol:** As teams build custom GPTs, create a central registry listing these AIs, their purpose, users, knowledge sources, and sharing level. This avoids duplication, promotes best practices, and gives your AI governance committee visibility.
- **Explore "Actions" for One Key System Integration to Enhance a Custom GPT:** With your IT team, identify one high-impact scenario where enabling a custom GPT to interact with another core hotel system via an "Action" (e.g., pulling PMS availability, logging a service request) could dramatically improve a workflow. Begin exploring technical feasibility, API availability, and security protocols for future advanced AI integration.

13
INTELLIGENT WORKFLOW AUTOMATION

FOR DECADES, HOTEL operations have been structured around human bottlenecks. Reservations pile up in inboxes. Requests move through layers of approval. Guest data sits locked in systems that fail to talk to each other. And your team? They are stuck in the middle, copying, pasting, reformatting, and manually pushing information through digital channels never designed to connect. We have accepted this as the cost of doing business in hospitality, a necessary friction in the path of your team's AI ascension toward higher-value work.

What if your digital systems could finally work together? What if your operation's routine, repetitive parts could happen automatically, freeing your team to focus on what humans do best: creating exceptional guest experiences? It is happening right now in hotels that have embraced intelligent workflow automation.

Beyond Templates and Chatbots: The Next Development of Hotel AI

In previous chapters, we have explored how to build AI literacy, master prompting, and create specialized custom GPTs customized to your hotel's unique needs. These were the foundational steps in your AI ascension. Yet, there is a fundamental limitation we need to address: all these approaches require someone to initiate the interaction. A human must remember to use the AI tool, craft the right prompt, and integrate the AI's output into their workflow.

Intelligent workflow automation flips this paradigm entirely. Instead of waiting for a human to ask, your AI systems become proactive participants in your hotel's operations automatically triggered by real-world events, seamlessly exchanging information across previously siloed platforms and acting precisely when needed. This involves eliminating the digital busy work that consumes their day, breaks their focus, and prevents them from delivering the high-touch service that defines exceptional hospitality.

What Exactly IS Automation in a Hotel Context?

Imagine a typical evening at your front desk when a new reservation comes in for a VIP guest. Today, that triggers a cascade of manual steps: The front desk agent notices the booking in the PMS. They copy the guest's information. They open your email system and find the appropriate welcome email template. They paste in the guest's details and personalize the message based on the guest's history before sending the email. They then switch to your team communication platform to notify the GM about the VIP arrival. They update the special requests log for housekeeping and flag the reservation for your F&B team to prepare a welcome amenity. That is easily fifteen to twenty minutes of purely administrative work multiplied by dozens of reservations daily. It is toil in its purest form: necessary and unfulfilling work that drains time and energy.

With intelligent workflow automation, everything changes. The moment that VIP reservation enters your system, it automatically pulls the guest's complete profile from your CRM. It sends this information to your connected LLM. The AI drafts a personalized welcome email in your exact brand voice, referencing the guest's preferences and past stays. A front desk supervisor gets a notification with the draft for quick review. The email is automatically sent once approved, taking seconds instead of minutes. Simultaneously, your GM receives a mobile alert about the VIP arrival. Housekeeping gets notified of the guest's room preference, and F&B receives instructions for the appropriate welcome amenity according to the guest's documented preferences. No one has to remember to check for new bookings. No one has to manually copy information between systems. No one has to draft repetitive emails. The trigger, a new VIP booking, sets everything automatically in motion. That is the essence of automation: connecting systems so they work together without constant human intervention.

Triggers, Actions, and the Digital Nervous System

The key to understanding automation is recognizing the relationship between triggers and actions. Triggers are specific events that kickstart the automated workflow. These can include a new booking in your reservation system, an incoming email with specific keywords, a guest checking in through your PMS, a room status changing to "clean" in your housekeeping system, a specific time of day being reached, a low inventory threshold being crossed, or a negative review being posted online.

Actions happen in response to those triggers. It is helpful to note that these automation actions are conceptually similar to the "Actions" capability you learned about in the custom GPTs section in Chapter 12, as both allow AI-driven systems to connect with other software and perform specific tasks driven by given inputs or events. Examples of automation actions include data being extracted and formatted, information being routed to the right team member, your LLM generating personalized content, updates being posted to your team chat, new records being created in your CRM, notifications being sent to guests or staff, or reports being generated and distributed.

The result happens when you connect these triggers and actions across different systems, creating a digital nervous system for your hotel that can sense important events and respond appropriately without constant human management. This is fundamentally different from prompting an AI. Prompting is when you ask a question and get an answer. Automation is setting up a system that says: "When X happens, do Y and Z automatically, then let humans handle the exceptions or high-value interactions."

The Technical Bridge: How Modern Tools Connect Your Digital Ecosystem

The hospitality technology environment has historically been fragmented. Your property management system (PMS) does not naturally talk to your customer relationship management (CRM), and your booking engine is disconnected from your team chat platform. This creates data silos and forces your staff into time-consuming manual work. Today's intelligent automation technologies solve this by acting as a powerful bridge between these disparate systems.

The solution involves three key types of technology working together. First are the automation platforms, such as Zapier, Make.com, n8n, or Microsoft Power Automate, which act as the central conductors for your workflows. Second, are the APIs (application programming interfaces), which are the standardized messengers that allow the automation platform to securely request and exchange information with your other software. The third, and what makes the process truly intelligent, is the large language model (LLM), which serves as the "brain" that can be called upon to perform a cognitive task.

Let's see how they work together in a practical example. Imagine a new VIP booking is made in your PMS. This event acts as the trigger for which the automation platform is listening. Once the trigger is detected, the platform begins a series of actions:

1. First, it uses the PMS's API to pull the guest's name, stay dates, and noted preferences.
2. Next, the platform sends this collected information to your specialized LLM with specific instructions, like "Draft a personalized pre-arrival welcome email based on this guest's data."
3. The LLM then analyzes the information, crafts the personalized email in your hotel's brand voice, and sends the completed draft back to the automation platform.
4. Finally, the platform performs the next action in the sequence, such as sending that drafted email to a front office manager for a quick review and approval before it is automatically sent to the guest.

This collaboration, where the automation platform guides the workflow, APIs transfer the data, and the LLM provides the custom intelligence, allows you to automate highly complex and personalized tasks. The real breakthrough is that this capability is more accessible than ever. Modern automation platforms offer visual, drag-and-drop interfaces that empower hotel leaders to design and implement these intelligent workflows without needing deep coding expertise or massive IT budgets.

From Robotic Process Automation to Genuine Intelligence

This new wave of intelligent automation dramatically differs from older approaches like traditional robotic process automation (RPA). Traditional RPA was essentially screen-scraping technology, robots programmed to click specific pixels on specific screens in a very rigid sequence. Change the interface slightly, and everything breaks down. It was automation without adaptation or understanding.

Today's intelligent automation, powered by large language models, brings actual cognitive capabilities to these workflows:

- **Natural Language Understanding**: It can interpret the content and intent of emails, reviews, or requests without just recognizing that they exist.
- **Contextual Awareness**: It understands the difference between a routine request and an urgent problem, adjusting its actions accordingly.
- **Content Generation**: It does more than just move information around; it can create personalized, on-brand responses, summaries, and recommendations.
- **Pattern Recognition**: It learns from data over time, identifying trends and anomalies that might escape human attention.
- **Judgment Within Parameters**: It can make rule-based decisions while escalating edge cases to human operators. This leap from mechanical repetition to contextual intelligence significantly changes ordinary automation into a true operational breakthrough for hotels.

Building Your Automation Toolkit

You do not need an enterprise IT budget or a team of developers to start implementing intelligent workflows. The tools have developed to be impressively accessible.

- Zapier is the easiest entry point for most hotels. It connects thousands of apps with simple if-this-then-that logic. For example, when a five-star review hits TripAdvisor, Zapier can automatically send it to your LLM, which drafts a response and then routes that draft to your marketing team for approval before posting. Zapier's strength is its vast library of pre-built connections to popular platforms, making it ideal for quick wins and basic workflows.

- Make.com is designed for more sophisticated, multi-step processes. Its visual builder handles complex logic, such as, "If a guest emails about early check-in, analyze the message sentiment with AI. If they are a loyal program member, automatically check room availability. If a suitable room is ready, confirm the early check-in. If a suitable room is unavailable, draft a personalized alternative option for staff review." Make.com shines when your processes need conditional branches, multiple data transformations, or parallel workflows.
- Microsoft Power Automate is the natural choice if your hotel is already invested in the Microsoft ecosystem. You can build flows that integrate seamlessly with Teams, SharePoint, Outlook, and Dynamics 365. When a corporate contract is up for renewal, Power Automate can trigger your LLM to analyze the client's history, draft a personalized renewal offer in Word, and notify the account manager via Teams. Its tight integration with Microsoft's security framework makes it particularly valuable for hotels with strict compliance requirements.
- n8n is for hotels with some in-house technical resources. This open-source platform gives you maximum control and data privacy, letting you host automation on your servers, which is crucial for properties with stringent data governance policies or unique integration needs. This provides unparalleled flexibility for custom integrations and workflows, although it requires a more technical setup.

The key is starting with the platform that best matches your current technical capabilities and most pressing operational needs. You can always expand or migrate as your automation strategy matures. Platforms like Zapier and Make.com often offer free or starter tiers for simple automation. As your workflows become more complex, involve more steps, or require premium app connections, you will likely need to move to paid subscription plans.

Reviewing their pricing tiers based on your anticipated usage to effectively budget is wise.

Human-Centered Design: The Critical Role of Human Oversight

Intelligent automation in hospitality aims to refocus human attention on where it adds the most value. The most successful hotel automations include human-in-the-loop (HITL). HITL strategically positions human oversight at key decision points. This includes reviewing guest communications before they are sent, especially for sensitive situations or VIP guests; approving financial transactions above certain thresholds or outside normal parameters; handling exceptions and unusual scenarios that fall outside the AI's training or authority; and providing final judgment on complex decisions with significant operational impact. This balanced approach removes repetitive drudgery while preserving human empathy, judgment, and accountability where they matter most.

Consider a service recovery workflow: When a negative review appears online, automation could instantly analyze the sentiment, identify the specific issues mentioned, pull the guest's stay history, and draft a personalized response. However, that draft should always be reviewed by a human manager before posting. Automation saves time by gathering all the relevant contexts and creating a starting point, but the final communication benefits from human empathy and judgment. This human-centered design principle should guide every automation you build.

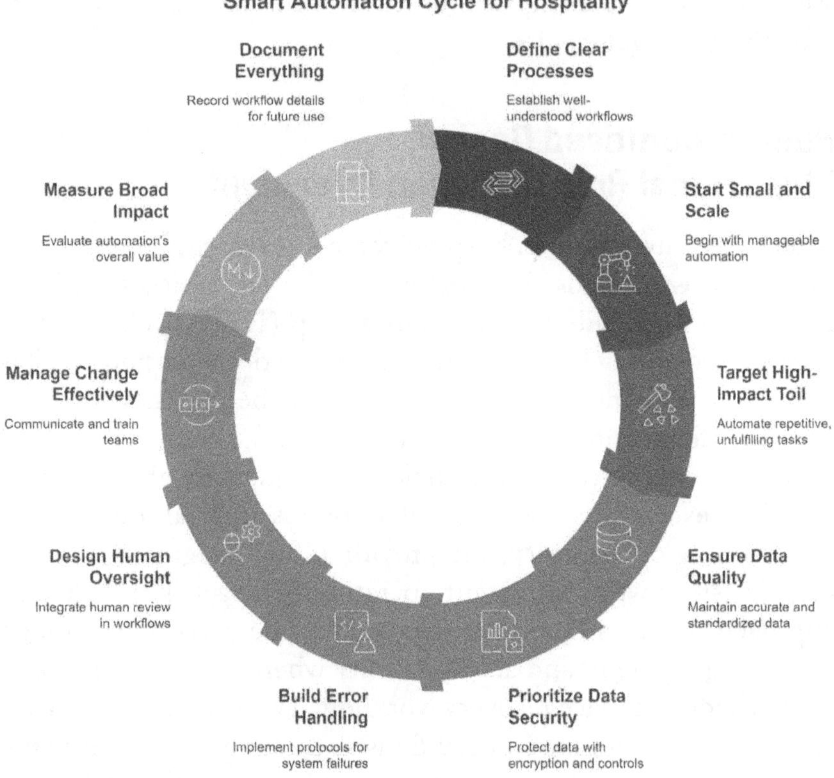

Smart Automation Principles for Hospitality Leaders

As you begin building your hotel's intelligent workflows, follow these principles:

- **Start With Clear, Well-Defined Processes**: Automation works best when the underlying process is already well-understood. Choose workflows with clear trigger points, predictable steps, and defined outcomes. Trying to automate chaos just creates faster chaos.
- **Begin Small, Then Scale**: Avoid attempting to automate your entire operation overnight. Start with one clear pain

point, perhaps automating responses to common email inquiries or streamlining your post-stay feedback process. Get it working smoothly, measure the impact, and then expand to adjacent processes.

- **Target High-Impact Toil First:** Remember the joy-toil analysis from Chapter 10? Look for tasks that score high on the toil scale: repetitive, rule-based, time-consuming, and deeply unfulfilling for your team. These are your prime automation candidates and will deliver the most immediate morale boost.
- **Focus on Data Quality and Context:** Your automation is only as good as the information flowing through it. Ensure that data moving between systems is accurate, standardized, and complete. Pay special attention to the context provided to your AI steps; an AI drafting a guest response needs full visibility into the guest's history and the specifics of their inquiry.
- **Prioritize Secure Data Handling:** Ensure that any data shared between systems via APIs in automated workflows adheres to your hotel's data security and privacy policies, as discussed in Chapter 8, with appropriate encryption and access controls.
- **Build Robust Error Handling:** Systems fail, APIs go down, and unexpected inputs appear; plan for these scenarios from the start. Every automation should have clear error handling protocols, such as automatic retries, notifications to the right people (e.g., if an automated pre-arrival email fails, the system could alert the front office manager via Teams with guest details and the error), fallback procedures, and comprehensive logging for troubleshooting.
- **Design for Appropriate Human-in-the-Loop Oversight:** For anything guest-facing or with significant operational impact, build in human review steps at the right points in the workflow. Your revenue manager should approve that dynamic pricing change before it goes live. Your guest

relations manager should review that service recovery email before it reaches the upset guest. Design these checkpoints to be efficient by presenting all the relevant context alongside the AI's recommendation to enable quick, informed decisions.

- **Manage the Human Element of Change:** Communicate clearly with your team about upcoming automation, explaining how they will support their roles rather than replace them. To ensure smooth adoption and buy-in, provide necessary training, address anxieties proactively, and involve them in the design process.
- **Define and Measure Broad Impact:** Beyond time savings, identify metrics like error reduction rates, improved guest satisfaction scores from quicker responses, higher staff morale from reduced toil, or increased conversion rates on automated offers to fully capture automation's value.
- **Document Everything:** As your automation ecosystem grows, documentation becomes vital. For each workflow, record its purpose, the systems it connects, its trigger conditions, the actions it takes, where human oversight occurs, and who to contact if something goes wrong. This documentation is invaluable for troubleshooting, onboarding new team members, and continuing to refine your processes.

Conceptual Examples: Intelligent Workflows in Action

Let us explore how intelligent automation can significantly change operations at forward-thinking hotels.

- **24/7 Personalized Booking Support**: Imagine a boutique hotel group implementing an intelligent workflow connecting their website inquiry form, their LLM, their PMS, and their sales team's Slack channel. When a potential guest submits a booking inquiry outside business hours, the system automatically analyzes the inquiry content, using the LLM, to determine the guest's needs; checks real-time availability in the PMS for the requested dates; uses the LLM to draft a personalized response with available options; sends this response to the guest within minutes, lacking hours of delay; creates a follow-up task for the sales team to personally connect during business hours; and logs the entire interaction in the CRM.

- **Intelligent Reputation Management:** Imagine a luxury resort created a workflow connecting its review monitoring platform, its PMS guest database, its LLM, and its team chat. When a new review appears on TripAdvisor or Google, the system automatically analyzes the sentiment and key topics. It pulls the guest's stay history and any incident reports from the PMS. The LLM drafts a personalized response according to this context. For positive reviews, the draft is sent to the marketing team for quick approval. The system alerts the department head and the GM for negative reviews. All draft responses include relevant policy details and past interactions.
- **Dynamic Upselling Based on Real-Time Inventory:** Imagine a city hotel built automation connecting their PMS, LLM, and email platform. The system checks current occupancy and available room upgrades three days before arrival. It analyzes the guest's booking history and preferences. The LLM generates a personalized upsell offer at an optimal price point. A pre-arrival email with the customized upgrade opportunity is automatically sent. Accepted upgrades are instantly processed in the PMS, and the revenue team receives a daily summary of all upsell conversions.
- **Intelligent Meeting and Event Management:** Imagine a conference hotel implementing a workflow connecting its event management system, its LLM, its catering system, and its operations platform. The system extracts all event

details and client requirements when a new event booking is confirmed. The LLM generates a preliminary banquet event order (BEO). Relevant departments like catering, AV, and housekeeping are automatically notified. The BEO draft is routed to an event manager for review and refinement. Upon approval, setup instructions are distributed to operational teams, and the client receives a personalized confirmation of the next steps.

These examples clearly illustrate that intelligent automation removes the friction that prevents your team from delivering exceptional service.

The Future of Hotel Operations is Hybrid

This development toward intelligent automation represents a fundamental shift in how hotels operate. We are moving from manual processes to intelligent workflows, from reactive service to proactive anticipation, from siloed information to seamless data flow, and from repetitive tasks to strategic focus. The future is AI plus humans, a hybrid workforce where digital systems handle the structured, repetitive processes while your team focuses on what makes hospitality special: genuine human connection, creative problem-solving, and memorable moments of service.

By building these intelligent workflows now, you are reshaping the very nature of hotel operations. You are freeing your team from digital drudgery and creating space for the work that brings real joy to them and your guests. That is the power of intelligent workflow automation: empowering people.

From Concept to Implementation

Are you ready to start building your first intelligent automation? Here is how to move forward:

- **Identify Your First Automation Target:** Choose a process that is frequently performed, relatively standardized, currently time-consuming, and low risk for initial implementation. Common starting points include automating standard email responses, pre-arrival communications, or review responses.
- **Map the Current Process in Detail**: Document how the process works today: triggers, systems, information flow, notifications, and human review points. This map is your automation blueprint.
- **Choose the Right Automation Platform**: Select the best platform, such as Zapier or Make.com, according to your technical capabilities and system connection needs. Zapier is often a good starting point.
- **Configure the AI Model (The LLM "Brain")**: This is where you design the intelligence of your workflow. Before building the full automation, create or instruct the specific AI model that will handle the core intelligent task, designed to draft personalized text, analyze sentiment, or categorize information based on the inputs it will receive from the automation's trigger.
- **Start Simple, Then Iterate**: Build a basic version connecting essential steps with HITL. Test thoroughly before a broader rollout, then add more sophistication.
- **Measure the Impact**: Establish clear metrics, such as time saved, faster responses, reduced errors, or team satisfaction. This data justifies further strategy investment.

The hotel groups that master this hybrid approach, blending AI efficiency with human expertise, will do more than save time and reduce costs. They will fundamentally change how work gets done, creating space for the moments that matter, those irreplaceable human interactions that define true hospitality. The automation revolution in hospitality is here. The question is: Will you lead it or be left playing catch-up?

Key Takeaways:

- Intelligent workflow automation connects disparate hotel systems and AI tools to enable proactive, event-triggered participation in operations, moving beyond human-initiated AI interactions.
- This automation connected to an LLM is facilitated by platforms like Zapier, Make.com, Microsoft Power Automate, and n8n, which use APIs to bridge different software, allowing for sophisticated, multi-step processes without extensive coding.
- Successful automation in hospitality is human-centered, with human-in-the-loop (HITL) oversight at critical decision points, ensuring AI augments human capabilities rather than replacing essential judgment and empathy.
- Effective implementation follows key principles: starting with clear processes, beginning small, targeting toil, ensuring data quality and security, robust error handling, strategic HITL design, managing change, measuring impact, and thorough documentation.

Bringing This into Focus:

- **Identify One Simple Automation Opportunity This Week:** Think of one task from your toil list (Chapter 10) that involves manually moving information between two systems, for example, from a booking confirmation email to a spreadsheet. Consider if a tool like Zapier could automate this if an AI first extracted the key data.
- **Explore an Automation Platform's Capabilities:** Visit the website of one of the automation platforms mentioned, such as Zapier or Make.com. Look at their "templates" or "integrations" page. See if you can identify connections relevant to your hotel's existing software. This is primarily for awareness building and requires little time.

- **Sketch a Basic "Trigger-Action" Flow:** Sketch a simple flow for the automation opportunity you identified in the first step. What is the specific trigger that would start the process? What is the first action an AI or system should take in response? What would be the desired outcome of this automated step? This helps visualize the automation logic.

Leadership Action Agenda:

- **Sponsor a Small Automation Pilot Project:** Based on your team's work from Chapter 3 regarding job deconstruction, select one well-defined, high-toil, low-risk process for a pilot automation project. Assign a small team, including an AI champion and someone comfortable with technology, to explore using a tool like Zapier or Make.com to automate it within the next thirty days, ensuring clear HITL checkpoints are designed into the workflow. Focus on measuring time savings and toil reduction as key outcomes.
- **Inventory Key Hotel Systems and API Availability for Automation:** Task your IT department or a tech-savvy team member with creating a simple inventory of your hotel's core software systems, including PMS, CRM, POS, guest messaging, and review platforms. Instruct them to investigate which of these systems offer APIs or have existing integrations with common automation platforms like Zapier or Make.com. This knowledge is foundational for planning more advanced intelligent automation.
- **Conduct a Human-in-the-Loop Design Workshop for Potential Automation:** Conduct a workshop with relevant stakeholders before automating any guest-facing or financially sensitive process. The goal is to collaboratively map out all necessary human-in-the-loop approval points and exception-handling procedures for the proposed automation. Document these HITL requirements clearly in your AI governance framework to ensure responsible implementation.

14 EMBRACING YOUR AI AGENTS

WE HAVE PROGRESSED significantly on our AI ascension journey, from understanding AI basics and mastering the art of prompting to creating specialized custom GPTs and leveraging intelligent, automated workflows. These automations, as explored in the previous chapter, represent a major leap in your hotel's efficiency, freeing up repetitive tasks. Each milestone is a key step in your AI ascension, empowering your team to work smarter and focus on higher-value contributions. Now, we stand ready to explore an even more significant frontier: AI agents.

An AI agent perceives its environment, reasons through information, and takes action toward achieving goals with minimal human intervention. Its architecture includes perception for gathering input, understanding for interpreting context, reasoning as its decision-making core, memory for maintaining consistency, and action for carrying out tasks via system integrations. This involves strategically empowering AI to take more initiative as you function like a trusted, proactive delegate on your operational team. The aim is to further reduce operational busy work and enable your human talent to dedicate their skills to strategy, creativity, and delivering exceptional guest engagement, elevating the meaningful aspects of their work and marking a new peak in your AI ascension.

A Note on AI Agents: Acknowledging the Leap

Before we more deeply examine the capabilities of AI agents, it is important to frame this stage of AI adoption realistically. Although the journey from creating custom GPTs, as explored in Chapter 12, and implementing intelligent workflow automation, detailed in Chapter 13, to embracing AI Agents is a logical progression, it also represents a notable increase in complexity and, often, the resources required for successful implementation. For the average hotelier, building, deploying, and managing sophisticated AI agents that can operate with significant autonomy and advanced reasoning is a more advanced undertaking. Although the tools for creating agents are becoming more accessible, effective and reliable agentic systems typically demand a greater investment in several areas. These include specialized skills, as designing and overseeing agents may require deeper technical understanding or partnership with AI specialists beyond general AI literacy. Significant time and rigor are needed because the design, training, and especially the testing and ethical oversight of autonomous agents require considerable time and careful attention to detail. There are also potential costs; platforms enabling sophisticated agent creation and integrating various hotel systems might involve higher subscription costs or development expenses. This chapter aims to provide a clear vision of what AI agents can achieve. Still, it is important to understand that mastering custom GPTs and workflow automation will be the primary focus for many in the near term. The "Practical Path Forward" section at the end of this chapter will offer tangible steps for beginning to explore agentic thinking. However, the full-scale deployment of complex AI agents should be approached as a strategic initiative, carefully aligned with your hotel's developing AI maturity and resource capacity. This understanding helps manage expectations and ensures a sustainable AI ascension.

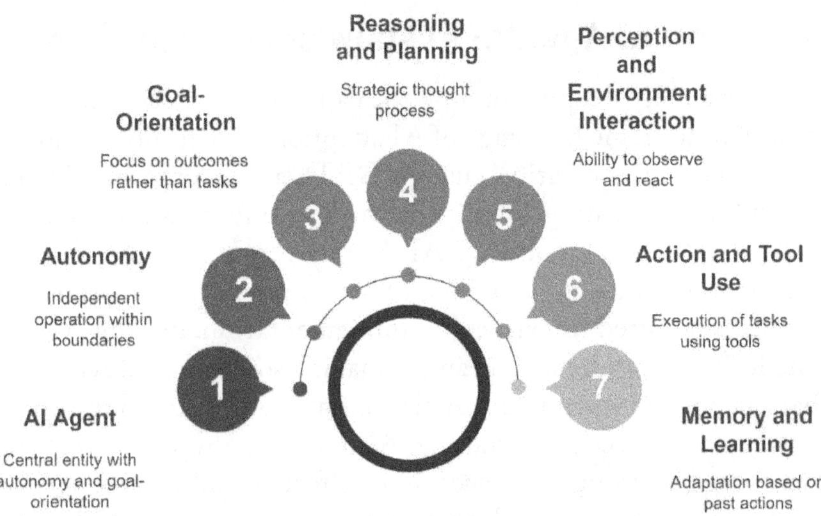

What IS an AI Agent? Defining Autonomy and Goal-Orientation

Unlike an automated workflow following an "if this happens, then do these steps" sequence, an AI agent operates with distinct characteristics.

- **Goal-Orientation**: Focusing on the what instead of just the how. You assign an AI agent a broader outcome-focused objective rather than discrete sequential tasks. Instead of automating steps to send pre-written offers, an agent might be tasked with: "Proactively identify past guests at risk of a failure to rebook according to stay patterns and feedback, then design and execute a personalized outreach campaign to encourage rebooking within the next quarter, aiming for a 10% conversion rate." The agent gets the what and figures out the how.

- **Autonomy:** Meaning they operate independently within boundaries for extended periods to achieve goals. They make certain decisions driven by programming, knowledge, and observations without requiring human intervention for each step. This autonomy always exists within carefully defined parameters and ethical guardrails set by human designers and managers.
- **Reasoning and Planning**: Which is strategic "thought." An agent can internally reason about its goal and environment. It breaks down complex assigned goals into logical sequences of necessary sub-tasks. It might evaluate different paths to achieve that goal and can adapt its plan if it encounters obstacles or new information.
- **Perception and Environment Interaction**: The ability to "observe." They can "perceive" their relevant digital environment, which could involve processing new data, like new bookings or competitor pricing changes, understanding user requests, or reacting to specific triggers, such as a weather alert impacting an event.
- **Action and Tool Use**: The ability to "do." They can autonomously access various digital tools to achieve objectives. These might include initiating web searches; querying your hotel's databases; interacting with APIs to check room availability, log CRM interactions, or send messages; or calling specialized AI models.
- **Memory and Learning for Contextual Continuity and Adaptation**: They maintain a more persistent "memory" of the situation and their goal than standard chatbots, allowing longer, more intricate problem-solving cycles. Some advanced systems can even "learn" from past actions, adjusting future behavior; this learning is often a form of adaptation according to performance data against set goals, usually within strict parameters and with human oversight

to ensure the agent avoids learning undesirable behaviors or deviating from ethical guidelines.

Think of it this way: An intelligent automated workflow is like giving a concierge a set of very specific, rigid instructions for a common task. It operates like a precise script. For example: "When a guest asks for a dinner reservation at 'the downtown steakhouse,' follow these steps: 1. Check the reservation system for 7:30 p.m. 2. If available, book it and send confirmation email #2. 3. If unavailable, check for 7:00 p.m. or 8:00 p.m. and offer those alternatives. 4. If those are also unavailable, respond with email #3." The workflow executes this predefined process flawlessly but cannot handle goals outside this narrow script.

An intelligent AI agent, on the other hand, is like giving an expert concierge a broader, more complex challenge. Instead of a script, you give it an objective: "Your goal is to design an unforgettable anniversary evening for the Miller couple in room 805. They enjoy classic Italian food and live jazz. Propose and secure all necessary reservations for them." The agent would then autonomously access different tools and data. It could check the couple's CRM profile for past dining preferences, search for top-rated local Italian restaurants with high sentiment scores, cross-reference that with nearby jazz clubs, check real-time availability for both, and then propose a complete, optimized itinerary with reservations for the human concierge to review and present to the guests.

You empower the agent with an objective, access to tools, relevant data, and the autonomy to create a comprehensive plan.

What AI Agents Lack

To maintain a clear perspective, it is equally important to understand the capabilities that current AI agents do not possess.

- **Sentience or Consciousness**: AI agents operate without feelings or self-awareness. Their "reasoning" is a form of sophisticated algorithmic processing, not genuine consciousness.

- **Infallibility**: AI agents are imperfect; they can make mistakes or misinterpret situations. Consequently, human oversight for handling exceptions and errors remains essential.
- **Strategic Accountability**: AI agents operate based on the goals and parameters set by humans, while the ultimate strategic direction, ethical judgment, and accountability always rest with human leaders.
- **True General Intelligence**: For now, current AI agents do not possess the capability of true general intelligence. They are typically specialized for specific tasks and cannot flexibly learn and apply knowledge across any intellectual field as a human can.

Understanding these distinctions is essential for setting realistic expectations and ensuring the beneficial integration of AI agents into your operations.

The "Mind" of the Agent: Frameworks like ReAct

How does an AI agent "think"? Conceptual frameworks help us understand. One influential framework is ReAct, an iterative cycle of reasoning and acting. Think of how you might decide what to wear: You observe the weather forecast, reason about your day's activities and clothing options, act by choosing an outfit, and then observe if you feel comfortable, adjusting if needed. AI agents use a similar iterative logic guided by ReAct.

First, they observe. The agent takes in the current state, such as new data, a user request, or an environmental trigger.

Then, they think/reason. This is its "internal deliberation." What is my goal? What info do I have or need? What are the potential next actions? Which tools help? Does it need human clarification or approval, especially if a human in the loop is designated for this decision?

Next, they act. The agent performs the chosen action, like drafting communication, querying your PMS, or escalating to a human.

Finally, they observe again. The agent assesses the outcome. Did it work? Has the situation changed?

This observe-reason-act-observe loop can repeat, allowing adaptive problem-solving. An agent might have internal modules for planning, tool use, and memory to enable autonomous behavior.

Agentic Flows: Coordinating Multiple AI Agents for Complex Hotel Objectives

The power of AI agents is magnified when they work together in agentic flows. Instead of one agent doing everything, an ecosystem of specialized agents collaborates to achieve larger objectives. Imagine a seamless guest arrival and personalized welcome agentic flow.

- A pre-arrival personalization agent would have the goal to ensure VIP/loyalty guest preferences are anticipated. It would observe by monitoring PMS arrivals. It would reason/act by querying the CRM for preferences; if found, for example, "hypoallergenic pillows," it tasks housekeeping or alerts guest relations. It uses an LLM to draft a personalized pre-arrival email for human review.
- A real-time arrival coordination agent would aim to ensure smooth check-in. It would observe by monitoring flight data, early check-in requests, and lobby traffic. It would reason/act by notifying the front desk if a VIP flight is delayed, alerting the FOM if queues form, and communicating with the pre-arrival agent for room readiness.
- An in-room welcome and service agent would focus on delivering an exceptional initial in-room experience. It would observe by receiving check-in confirmation. It would reason/act by triggering room service for a welcome amenity,

activating personalized in-room tablet settings, and routing in-room voice requests.

- A post-interaction feedback and follow-up agent would aim to gather feedback and address service issues. It would observe by monitoring service system logs. It would reason/act by triggering personalized SMS for feedback, and if negative sentiment is detected, it escalates to a human guest service manager. Each agent has a distinct role in this flow, sharing information to create a superior guest experience, with human-in-the-loop oversight ensuring alignment with hotel standards.

Building and Deploying AI Agents in Your Hotel: Tools, Considerations, and the Human Role

Tools for building sophisticated agents are becoming more accessible, often using no-code/low-code principles. Platforms help create and manage agents through visual interfaces. Successfully building and managing agents requires careful strategy, unwavering human oversight, and an understanding of resource implications. Deploying AI agents can involve specific costs, including subscriptions for agent-building platforms, fees for API access to various systems your agent needs to interact with, and potential costs associated with the agent's computational resources. Developing and maintaining advanced agents might also require specialized skills, which could involve training existing staff or engaging external consultants for complex setups.

Key considerations include:

- Clear goal definition by precisely defining the SMART objective for each agent.
- Defined operational boundaries regarding the absolute dos and don'ts and which decisions are autonomous versus which escalate to a human in the loop.

- Access to robust, reliable, high-quality data because agent decisions depend on data quality, so ensure accurate, up-to-date, well-structured data is securely accessible.
- Secure tool access protocols using the principle of least privilege by granting agents only the minimum necessary permissions.
- A commitment to continuous monitoring and refinement is necessary because deploying an agent is not setting it and forgetting it, so humans must track performance, identify errors, and refine instructions. Integrating agents with older hotel legacy systems lacking modern APIs can also pose challenges, potentially requiring custom IT work.

Essential Ethical Guardrails for AI Agents

The increased autonomy of AI agents requires more oversight.

- **Reinforce Human Accountability**: Human designers, managers, and your AI governance committee are ultimately accountable for agent actions and their consequences.
- **Demand Transparency and Auditability**: Prioritize agent platforms that clearly log agent decisions, data sources consulted, and actions taken. This allows for auditing if issues or unexpected outcomes arise.
- **Proactive Bias Monitoring**: Implement processes for humans to regularly review agent performance and decision patterns for any signs of amplified or emergent bias, especially if agents "learn" or adapt over time.
- **Strict Scope Adherence and Escalation**: Design agents with clearly defined operational boundaries. Implement mechanisms to prevent scope creep and ensure agents reliably escalate situations or decisions outside their mandate to designated human personnel.

Strategic Applications

Identify recurring, complex, multi-step processes where an agent's reasoning, tool use, and sustained operation can deliver significant value.

- **An AI sales lead management and qualification agent could have the goal:** "Ensure every inbound group lead is responded to within thirty minutes, fully qualified, and routed to the best sales manager with a briefing." It would monitor lead channels, use AI for qualification, research prospects, draft emails for human review, log CRM details, and notify sales managers, freeing their time for relationship building.
- **An AI guest service and online reputation management agent could aim to**: "Monitor all online channels for hotel mentions and facilitate timely, empathetic, brand-aligned responses." It would scan platforms and use sentiment analysis. For positive reviews, it drafts thank-yous for human approval. For negative feedback, it drafts responses and proposes recovery steps for human review. It summarizes, gathers guest history, and escalates to management for severe complaints.
- **A new staff onboarding and continuous learning agent could have the goal:** "Ensure every new team member feels welcomed, is fully trained on core competencies, and successfully integrated within thirty days." It could deliver personalized learning, answer common questions from SOPs, schedule meetings, send reminders, and flag to HR if someone needs extra human support.
- **An AI daily operational and financial reporting agent might be tasked to:** "Provide all key leaders with a concise, actionable summary of yesterday's performance and critical alerts by 7:00 a.m. daily." It would access PMS, POS, guest feedback, and labor systems data, analyze with specialized custom AIs, and compile a flash report.

The Developing Partnership: Humans Skillfully Managing, Guiding, and Collaborating with AI Agents

AI agents are designed to augment your human team. Even advanced AI agents require thoughtful human involvement in their design, training, ongoing monitoring, quality control, strategic direction, and ethical oversight. As AI agents take on more operational tasks, your team's roles will naturally develop, becoming more strategic. This shift includes responsibilities such as:

- **Designing and Refining Agents:** Humans will design, configure, and continually refine AI agents to meet specific hotel needs. This includes training and retraining agents by providing new information or adjusting their instructions.
- **Strategic Supervision (The Agent Boss Role):** A core function will be strategically supervising agent performance and ensuring agents operate in alignment with hotel objectives. This agent boss role involves:
    - Regularly reviewing agent decision logs and audit trails to ensure they perform as intended and adhere to ethical guidelines.
    - Defining, monitoring, and refining key performance indicators (KPIs) for agents based on their assigned goals.
    - Iteratively refining agent instructions and operational boundaries based on performance data and evolving hotel needs.
- **Ethical Governance and Control:** Humans must ensure ethical compliance in all agent operations, particularly concerning bias, fairness, and data privacy. They serve as the designated human in the loop, ready to intervene and override agent actions when necessary.

- **Troubleshooting and Exception Management:** Identifying and addressing agent errors and managing exceptions or situations escalated by the agent are key human responsibilities.

Beyond direct agent management, human team members will increasingly focus on tasks that AI cannot replicate. This includes expertly handling complex exceptions that require deep human empathy, driving continuous improvement in human-AI collaboration, and concentrating on high-value human interactions. With AI agents managing routine tasks, your team is freed for more strategic thinking and building strong guest relationships.

Full automation of high-risk or ethically sensitive processes should be cautiously approached. Any process involving significant financial risk, deep ethical considerations, guest distress, or requiring genuine human empathy must always retain a skilled, accountable human in the loop. Remember, AI agents are powerful co-pilots; they are not intended to be sole decision-makers.

The Journey Ahead: AI Agents as Digital Teammates, Coordinated by Human Wisdom

You are building a new type of digital workforce when you design AI agents with clear goals, provide them with knowledge, establish unwavering ethical boundaries, and ensure their actions are guided by skilled human judgment. A critical feature of these AI agents is their capacity to operate 24/7/365; they do not require breaks, observe weekends, or take vacations. This constant operational availability offers significant power. Furthermore, many agents are designed to learn and improve as they process more information and execute more tasks. These agents can be diligent, consistent, and data-savvy, capable of handling complex tasks at a scale and speed previously unattainable.

This capability strategically frees your human team to focus their expertise, creativity, empathy, and insights on the highest-value contributions. These include visionary leadership, creative innovation, detailed problem-solving, and the art of delivering exceptional hospitality. Achieving this vision requires mature AI literacy across your organization, a commitment to continuous learning, and an unwavering focus on ensuring these digital "teammates" fully align with your hotel's core values and ethical standards.

The Practical Path to Your First AI Agent

To move from concept to reality, consider this methodical approach:

1. **Assess Your Foundation**: Begin by evaluating your current automation infrastructure. Your existing custom GPTs, workflow tools, and system integrations serve as the essential building blocks for more advanced AI agents.
2. **Identify a Promising Pilot Use Case**: Select a process that is complex yet well-defined, time-consuming, standardized, involves multiple systems or data sources, and has clear metrics for success.
3. **Map the Process in Detail**: Document the current state of this chosen process, noting how it works today, its decision points, the systems involved, and existing human touchpoints. This detailed map will form the blueprint for your first AI agent.
4. **Define Clear Boundaries**: Determine which decisions the agent will handle autonomously and which require human approval or intervention.

Starting Small, Thinking Big: Your AI Agent Development Path

Successful AI agent implementation is rarely an overnight transformation. Instead, it is an iterative journey:

- **Begin with a Focused Agent**: Start by deploying a single agent focused on one well-defined process, or even just a part, where success can be clearly measured.

- **Prove, Learn, and Expand:** Use this initial deployment to prove the concept, gather learnings, and then methodically expand its capabilities or apply the model to other areas.
- **Document Everything:** Keep detailed records of what works, what challenges arise, where human intervention proves essential, and any unexpected benefits. This documentation becomes a valuable template for broader implementation. As your team gains comfort and experience, opportunities for leveraging AI agents will become apparent across your operation.
- **Maintain Balance:** The goal is to automate sensibly while always enhancing the crucial human elements of your service.

The Human Advantage and Strategic Impact

As AI agents take on greater operational complexity, the uniquely human aspects of hospitality become even more prominent. AI frees your team to more fully express genuine empathy, engage in creative problem-solving, apply intuitive understanding, and provide inspiring leadership. When intelligent systems handle routine processes, your people can dedicate their energy to creating memorable human connections. This is the essence of the AI-augmented hotel: technology managing complexity so humans can focus on the art of the guest experience.

Hotels that begin embracing AI agents now will develop valuable institutional knowledge about effective human-AI collaboration. Their teams learn to think differently, leading to more consistent service and meaningful guest interactions. This fosters a virtuous cycle: AI handles complexity, freeing humans to deliver exceptional experiences, which builds loyalty, drives revenue, and funds further innovation. The hotels that thrive in the coming years will be those that skillfully blend technological capability with human oversight, using AI agents as powerful force multipliers.

AI agents represent the next horizon in hospitality, amplifying the human touch by managing complexity behind the scenes.

Mastering this balance will set new standards for operational excellence and guest experience. Your journey begins with the first agent you choose to deploy. While the technology will continue to evolve, the guiding principle remains constant: Use AI to manage complexity so your team can focus on connection. This is how technology can best serve hospitality's highest purpose.

This chapter introduced AI agents as AI systems with greater autonomy, goal orientation, planning/reasoning, tool use, and adaptation capabilities, representing the next step in your AI ascension after intelligent automation. We clarified what AI agents are, driven by goals and using frameworks like ReAct to observe-reason-act, and what they lack, such as sentience or infallibility, emphasizing that they are not replacements for human strategy. We explored how multiple specialized agents can collaborate in agentic flows to achieve complex hotel objectives. Building and deploying agents requires clear goals and governance, including specific ethical guardrails for agents and always a human-in-the-loop (HITL) for oversight and critical decisions.

Key Takeaways:

- AI agents represent an advanced stage of AI, characterized by goal orientation, autonomy within defined boundaries, reasoning and planning capabilities, perception of their digital environment, and the ability to use tools and maintain memory to achieve objectives.
- Understanding agentic frameworks like ReAct (observe-reason-act-observe) and the potential of agentic flows (multiple specialized agents collaborating) is key to envisioning their application for complex hotel objectives.
- Building and deploying AI agents, while becoming more accessible, requires careful strategic planning, robust ethical guardrails, significant human oversight (HITL), and consideration of resource implications, including potential costs and specialized skills.

- The human role evolves toward designing, managing (agent boss), and collaborating with AI agents, focusing on strategic direction, exception handling, and tasks requiring deep empathy and complex judgment, ensuring AI augments rather than replaces human expertise.

Bringing This into Focus:

- **Distinguish Automation from Agentic Thinking for One Workflow:** Review one automated workflow your hotel currently has or is considering. Now, reimagine it: If you were to assign an AI agent to achieve the overall goal of that workflow, how might its approach and actions differ from the fixed steps of the automation? What specific tools or reasoning capabilities would it need to effectively pursue that goal with more autonomy?
- **Identify One Potential Agentic Goal in Your Hotel**: Think of one complex, recurring hotel challenge or objective that currently consumes significant human time and coordination, for example, "ensuring consistent brand messaging and service standards across all guest touchpoints and communication channels" or "proactively identifying and mitigating potential service failures for high-value VIP guests before they occur." Could an AI agent assigned with this clearly defined goal and provided with appropriate tools and data access contribute meaningfully to achieving this objective?
- **Brainstorm Critical HITL Points for an Agent Concept:** For the agentic goal you identified in the previous step, collaboratively brainstorm two to three critical points in the process where a human in the loop would absolutely need to be involved for review, approval, or final decision-making if an AI agent were autonomously working toward that goal. Document why human intervention is vital at these specific junctures.

Leadership Action Agenda:

- **Initiate an Intro to AI Agents Discussion with Your Leadership/Governance Team:** Share the core concepts of this chapter, including what AI agents are, how they differ from standard automation, the ReAct framework, the potential of agentic flows, and the essential ethical guardrails, with your key leaders and AI governance committee. Facilitate a discussion around one high-level hotel objective where an AI agent might offer significant operational value or competitive advantage in the future and with careful planning. This initial conversation plants the seed for future strategic thinking.
- **Task an Exploratory Review of Agent-Building Platforms:** Assign your IT team or a tech-savvy AI champion the task of researching one or two accessible platforms that facilitate building AI agents. Request a brief overview of their capabilities, potential ease of use for non-developers, integration possibilities with existing hotel systems, and indicative cost structures purely for educational and future planning purposes at this stage.
- **Define No-Go Zones for AI Agent Autonomy Within AI Governance:** As part of your next AI governance committee review, as discussed in Chapter 8, explicitly discuss and document specific no-go zones or critical decision types where fully autonomous AI agent action would be unacceptable for your hotel. Clearly define situations where direct, mandatory human-in-the-loop approval will always be required, for example, final hiring decisions, major unbudgeted capital expenditures, or responses to severe guest safety or security incidents. This proactive boundary setting is crucial before considering deploying more autonomous AI systems.

15
THE AI NEXUS HOTEL

WE ARE ENTERING a new reality in which AI can reason and solve problems. This readily available "intelligence on tap" is set to rewrite the rules of business and significantly change knowledge work as we know it. As highlighted in Microsoft's 2025 Work Trend Index Annual Report, we are seeing the birth of the "Frontier Firm": an entirely new kind of organization structured around on-demand intelligence and powered by hybrid teams of humans and AI agents. This signals the emergence of a changed hotel model for the hospitality industry. Although Microsoft frames this development broadly, for this playbook, we have termed this advanced, AI-integrated entity The AI Nexus Hotel.

The AI Nexus Hotel: Clarifying the Vision

It's natural to have questions about what the AI Nexus Hotel truly is. Is it a destination? An abstract ideal? The answer is that it's best understood as an operating philosophy and a strategic North Star rather than a static endpoint. Technology will always evolve, so the goal is not to achieve a final state but to build an organization with the culture, skills, and infrastructure to continuously operate in this AI-native way. It is a concept *and* an ideal. A model your hotel can strive for to remain competitive and deliver exceptional experiences.

For a hotel to be considered an AI Nexus Hotel, it does not mean every single department must use AI agents for every task simultaneously. Rather, it means the hotel's foundational structure has been rebuilt around AI. Data flows seamlessly between systems, and the *capability* for intelligent automation and agentic work exists across the organization, even if different departments are at different stages of their AI ascension. AI automation is not the end goal but a critical *component* of becoming an AI Nexus Hotel. The full vision also encompasses the human elements of ethical governance, AI literacy, and a transformed workforce.

Putting It All Together: The Journey to the AI Nexus Hotel

This entire playbook has been a systematic journey, providing the building blocks to construct your AI Nexus Hotel. Each chapter laid a crucial part of the foundation.

- **The Foundation:** We began by establishing AI literacy and a proactive AI mindset, understanding that you cannot build a new type of organization without first teaching your team how to think and speak the language of AI.
- **The Strategy:** We then moved to strategy, learning to use job deconstruction and the joy-toil framework to identify where AI could be most impactful. You learned to master your AI toolkit by understanding the different models, how to prompt them effectively, and how to use them to create everything from documents to your brand's unique sights and sounds.

- **The Guardrails:** We established the non-negotiable guardrails with the strategy in place. An AI Nexus Hotel cannot exist without robust ethical governance and a culture of critical thinking, ensuring that human judgment always guides the technology.
- **The Engine:** Finally, we built the operational engine. We learned how to create specialized custom GPTs, connect LLMs through intelligent workflow automation, and ultimately deploy AI agents to create a proactive, intelligent system.

All these pieces come together in the AI Nexus Hotel. It is an organization where AI acts as the intelligent connective tissue, creating a responsive and deeply interconnected ecosystem. It is powered by a work chart of human-agent teams, where every employee is empowered to be an agent boss responsible for their agent's performance. This practical blueprint based on technologies available today shows how your hotel can use AI to amplify human capabilities, embrace agile structures, and uphold a clear, humanity-first, AI-powered ethos. It is the ultimate expression of using technology to reduce toil, create more time for meaningful work, and elevate the joy of guest service and employee experience.

Charting Your Hotel's Journey to an AI-Integrated Future

Understanding how a hotel develops into an AI Nexus Hotel involves recognizing that this is a journey. Microsoft provides a valuable lens for this development with its concept of the Frontier Firm, seeing organizations progress through three distinct phases as they deeply integrate AI.

Microsoft's journey begins with Phase 1, where AI steps in as a helpful **assistant**. In this foundational stage, the primary aim is to lighten the daily workload. AI tools are adopted to smooth out the rough edges of existing tasks, helping your team members accomplish their current responsibilities more efficiently, effectively tackling routine drudgery.

As your hotel becomes more adept, it moves into Phase 2, welcoming AI agents as **digital colleagues**. Here, these AI entities become active members of your teams, capable of taking on specific tasks. This concerns equipping your employees with new AI collaboration skills, allowing them to significantly scale their impact and delegate parts of their workload to their digital counterparts.

The most advanced stage in Microsoft's vision is Phase 3, where **humans set the direction for agent-run processes**. At this point, AI agents can manage entire business processes. Human oversight shifts to a more strategic role: setting the overarching goals, monitoring the outcomes, and intervening when necessary to manage exceptions or refine strategies. This reflects a mature ecosystem where AI operates with a significant degree of autonomy on defined processes.

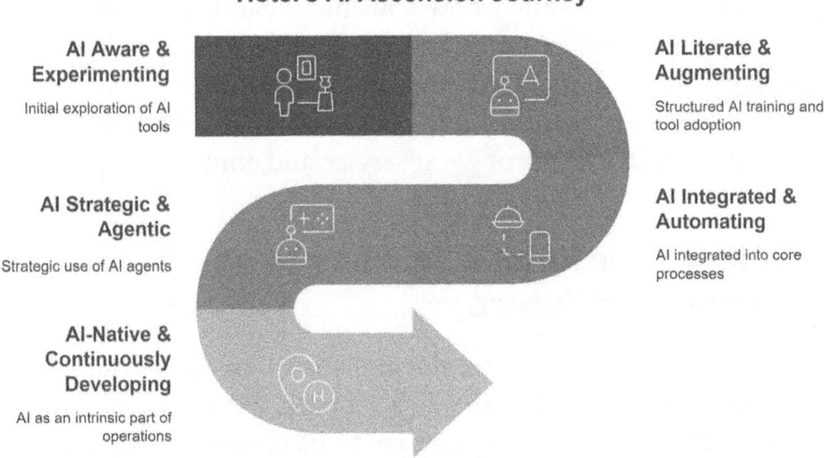

To bring this broad framework into sharper focus for the hospitality industry, our AI ascension levels offer a more detailed roadmap, charting your hotel's specific path and mapping directly to these Microsoft phases.

- Your hotel's journey often begins at AI Ascension Level 1: **Exploration**, a phase of grassroots curiosity where individuals might informally explore public AI tools for isolated tasks. There is an initial spark, but formal strategy or governance is still to emerge, meaning the impact on operations is minimal.
- This naturally progresses to Level 2: **Augmentation**, where the organization actively invests in building foundational AI literacy through structured training. Here, your staff consistently uses approved AI tools to boost their productivity, perhaps drafting initial email responses or brainstorming ideas for guest services. You might see the first custom GPTs emerging, and initial ethical guidelines start to take shape, focusing on enhancing tasks and saving pockets of time for your team. Levels 1 and 2 firmly align with Microsoft's Phase 1 (**AI as an Assistant**).
- As your hotel advances, you will reach AI Ascension Level 3: **Integration**, which, along with the initial stages of Level 4, corresponds to Microsoft's Phase 2 (Agents as **Digital Colleagues**). At Level 3, AI becomes integrated into your core hotel processes. Intelligent workflow automation begins to bridge the gaps between AI capabilities and your key systems, like the PMS or CRM, streamlining complex workflows. Custom GPTs have become more widely used and refined, and an active AI governance committee oversees ethical and effective deployment. The results become tangible here, with measurable improvements in operational efficiency and significant time savings.
- This leads into the early stage of Level 4: **Strategic and Agentic**, where your hotel strategically employs AI agents for more complex, goal-oriented tasks. Think of AI agents assisting with aspects of lead qualification for group sales or managing personalized pre-arrival communication sequences, always guided by human oversight.

- Finally, the later stage of Level 4 and the entirety of Level 5: **AI-Native** (The AI Nexus Hotel) represent the culmination of this journey, aligning with Microsoft's Phase 3 **(Humans Set Direction for Agent-Run Processes)**. In late Level 4, AI-driven data analytics become central to informing your hotel's strategic decisions. Job roles naturally reflect a true human-AI collaboration, with the human-in-the-loop principle being a well-understood and practiced standard. Reaching Level 5 means your hotel operates as an AI Nexus Hotel. Here, AI is an intrinsic part of your operational DNA. Organizational structures become agile, designed around dynamic human-AI teams. Data flows seamlessly to inform human and AI decision-making, and AI ethics are deeply embedded in the culture and process. Your hotel becomes a learning organization, continuously adapting and leveraging emerging AI capabilities. Imagine your guest relations manager, no longer bogged down by manual report generation, instead starting her day by reviewing AI-surfaced opportunities to create truly unique surprise-and-delight moments for arriving guests, co-creating these exceptional experiences with an AI assistant that has a deep understanding of individual preferences and real-time hotel resource availability.

Understanding these distinct levels and their correspondence to broader industry phases helps you chart a clear path for your hotel's AI ascension, allowing you to celebrate progress and strategically identify the next steps. It is also important to recognize that many organizations and hotels will operate across all three of Microsoft's phases simultaneously, with different departments or functions progressing at their unique pace.

Reimagined Hotel Organizational Structures in the AI Nexus Hotel: The Rise of the Work Chart

The AI Nexus Hotel operates with structures vastly different from traditional, siloed models—with AI providing intelligence on tap as new agile ways of working emerge.

- **The Work Chart and Cross-Functional, Human-Agent Teams**: Instead of rigid departmental silos, the AI Nexus Hotel organizes work around goals or guest journey stages, forming dynamic, outcome-driven teams. This concept, which Microsoft calls a work chart, replaces the traditional org chart. These cross-functional teams blend human expertise from various departments with AI agents that expand employee scope and enable faster more impactful work.
- **Flatter Hierarchies and Empowered Agent Bosses**: As AI delivers real-time data and decision support, many middle management layers focused on information relay become less necessary. Every employee, from the front line to the boardroom, has the potential to become an agent boss, building, delegating, and managing agents to amplify their impact. A guest service agent, guided by an AI recommendation engine, might resolve issues on the spot without multiple approvals.
- **Centralized AI Strategy, Decentralized Innovation, and Intelligence Resources**: Overall AI strategy and ethical governance are managed centrally by an AI governance committee. However, AI application and innovation are decentralized, with teams and AI champions encouraged to develop solutions within the governance framework. To manage this developing situation of human-agent collaboration and decentralized innovation, AI Nexus Hotels may see the rise of a new pivotal function: an Intelligence Resources Department. Conceptualized as a strategic blend of traditional IT and human resources capabilities, this

department would be dedicated to enabling and optimizing the hotel's AI ecosystem and its interaction with human talent. Its responsibilities would extend beyond managing the optimal balance of human and digital labor to include critical support functions, such as:

  - *AI Tool and Platform Governance Support*: Assisting in identifying, evaluating, and onboarding new AI tools under the direction of the AI Governance Committee (see Chapter 8).
  - *Centralized Prompt Library Curation and Best Practices*: Curating and maintaining a central library of approved, high-performing prompts (see Chapter 5).
  - *Facilitating AI Skill Development and Cross-Departmental Knowledge Sharing*: Partnering with L&D or leading AI upskilling and supporting the AI champion network (see Chapter 8).
  - *Supporting Human-Agent Team Effectiveness*: Providing guidance to agent bosses on managing AI agents and optimizing workflows.
  - *Change Management and AI Adoption Advocacy*: Communicating AI benefits and addressing concerns.
  - *Monitoring AI Impact and Value Realization*: Assisting departments in tracking metrics to ensure AI initiatives deliver tangible value. This intelligence resources function acts as a crucial enabler.

- **Fluid Roles and the Human-Agent Ratio:** The traditional, static job description becomes obsolete in an AI-augmented hotel. It is replaced by a more dynamic and agile approach to talent, centered on two key concepts: fluid roles and the human-agent ratio.

- **Fluid Roles: Shifting from Tasks to Skills**

  Traditionally, hotel roles have been defined by a fixed list of tasks; for example, a front desk agent checks guests in, and a reservations agent takes bookings. In the AI Nexus Hotel, where AI agents and automation handle many of these routine, procedural tasks, a person's value shifts from the specific tasks they perform to the unique human skills they contribute. Roles become fluid, meaning a team member's responsibilities can adapt based on the needs of a specific project or guest-centric goal.

  For instance, an employee who was previously a reservations agent might now contribute their persuasive communication and complex problem-solving skills to a temporary high-value group conversion team. They would work alongside an AI agent that handles the initial lead qualification and data gathering. However, humans build relationships, understand nuanced needs, and close complex bookings. This approach makes the organization far more agile, allowing leaders to assemble the best possible team of human skills and AI capabilities for any given challenge rather than being constrained by rigid departmental silos.
- **The Human-Agent Ratio: A New Metric for Leadership**

  To manage these new fluid teams, a new key metric emerges: the human-agent ratio. This is not a simple measure of staff versus bots in the hotel; rather, it is a strategic calculation for a specific task or project to optimize the balance of human oversight with AI efficiency. As an agent boss, a leader must deliberately decide on the right ratio for any given workflow.

Consider these two examples:

- **Low Human-Agent Ratio (High AI Involvement):** The ratio might be very low for a task like analyzing ten thousand guest reviews to identify common themes. One human manager might oversee an AI agent that performs 95% of the initial data processing and sentiment tagging. The human's role is high value: to interpret the strategic insights from the AI's analysis, not to do the manual work.

- **High Human-Agent Ratio (High Human Involvement):** Conversely, the ratio would be very high for managing a delicate service recovery situation with a distressed VIP guest. An AI agent might perform 5% of the task by instantly gathering the guest's complete history and summarizing the issue for the manager. The human manager, however, handles 95% of the work: applying empathy, building rapport, engaging in creative problem-solving, and making the final decision.

  This metric is important because it forces leadership to be intentional about where to apply their most valuable resource: their team's talent. It shifts the mindset from "What can we automate?" to "What is the optimal blend of human and AI collaboration to achieve the best possible outcome?"

New Economic Models and Value Drivers in the AI Nexus Hotel

AI integration reshapes the AI Nexus Hotel's economic model.

- **Hyper-Personalization as a Core Revenue Driver:** AI enables unprecedented guest understanding and service personalization at scale, moving beyond room revenue to maximize total guest lifetime value.
- **Radical Operational Efficiency and Addressing the Capacity Gap:** Intelligent automation and AI-driven predictive analytics optimize every resource, including staff time, energy, and inventory. This involves leveraging digital labor to close the capacity gap, the deficit between rising business demands and the sustainable output of humans

alone. Microsoft data reveals that 53% of leaders say productivity must increase, while 80% of the global workforce report lacking time or energy. AI helps bridge this.

- **Data as a Strategic Asset and Intelligence on Tap:** The AI Nexus Hotel treats ethically managed data as a core asset. AI unlocks deep insights, or intelligence on tap, for improving operations and innovating new AI-driven services or revenue streams. For example, analyzing guest data patterns might lead to an AI-powered hyper-personalized weekend itinerary planner as a premium ancillary service.
- **Enhanced Agility and Responsiveness to Market Dynamics:** AI-powered market intelligence and demand forecasting give the AI Nexus Hotel an edge in adapting rapidly to changing market conditions.

Everyone's an Agent Boss

As AI integration deepens within the AI Nexus Hotel, the nature of human work transforms. The most important development is the emergence of a new role and mindset that applies to every team member: the agent boss. This means every employee, from the front desk to the executive office, will be responsible for directing, managing, and collaborating with a portfolio of specialized AI agents to accomplish their operational and strategic goals. This shift from simply *using* AI tools to actively *managing* digital labor empowers every individual to amplify their impact across the hotel.

What Does Being an Agent Boss Mean for Daily Work?

Being an agent boss fundamentally changes the what and how of daily tasks. It's about moving from a doer of repetitive work to a director of an AI that handles that work.

- **For a Front Desk Agent:** Instead of manually typing out every response to a common question, they act as the agent boss for a guest communications agent. They oversee the AI, review its drafted responses for empathy and accuracy in complex situations, and empower it to handle routine inquiries autonomously, freeing up their time to provide high-touch, personal service to the guests standing before them.

- **For a Sales Manager:** Instead of spending hours pulling lead data and creating basic proposals, they direct a sales analyst agent. They set the goal—for example, "Find five new potential corporate clients in the biotech industry"—and the agent performs the research and initial data compilation. The sales manager then uses that output to focus on the human skills of building relationships and strategic negotiation.

Why is This Shift to Agent Boss Important?

This new role is critical because it's the mechanism for turning AI's potential into real-world productivity and enhanced guest experiences. It allows your hotel to scale personalized service and operational efficiency in a way that is impossible with human labor alone. When every employee can leverage a team of tireless, 24/7 AI agents, they are no longer limited by the number of hours in a day. Their value is amplified; they can achieve more, think more strategically, and are freed from the toil that leads to burnout, allowing them to focus on the joy of creative problem-solving and genuine human connection.

What Does This Mean for the Organizational Chart and Leadership?

The rise of the agent boss flattens traditional hierarchies. The org chart becomes less about a rigid chain of command and more about a dynamic work chart, where human-agent teams are formed to tackle specific goals.

- **For the Organization:** This creates a more agile and responsive structure. Instead of information moving slowly up and down a chain of command, AI agents can deliver relevant data directly to the people who need it, enabling faster decision-making at all levels.
- **For Leaders:** A general manager or department head becomes a super-agent boss. Their primary role shifts from managing their team's day-to-day tasks to setting the strategic vision for their department's portfolio of AI agents. They are responsible for coaching their team on how to be effective agent bosses, ensuring ethical AI use, and focusing on high-level strategy, human development, and managing key relationships.

What Does This Mean for Someone Starting Their First Job?

This concept is perhaps most transformative for those new to the workforce. In a traditional model, entry-level employees often spend their first year on repetitive, low-impact tasks. In the AI Nexus Hotel, a new hire is an agent boss from day one. They are immediately entrusted with managing AI agents to handle routine work, allowing them to contribute to more significant, strategic projects much earlier in their careers. Their job is to learn how to effectively direct their AI teammates to achieve the best results, making their role more engaging and accelerating their professional growth.

Overcoming the Hurdles: Navigating the Path to Becoming an AI Nexus Hotel

Microsoft's "Path Forward" offers guidance: 1. Hire your first digital employees (agents for valued tasks). 2. Set your human-agent ratio (invest in people, identify automation/collaboration points).

3. Get to a broad scale fast (move beyond pilots, target high-need areas, reinvest value).

The journey to an AI Nexus Hotel has challenges. These include:

- The need for leadership vision and commitment requires a strong championship (the tone from the top, as discussed in Chapter 2).
- Organizational culture change involves embracing human-AI collaboration and is addressed by fostering AI literacy and mindset, as detailed in Chapter 2.
- Developing new skills and mindsets by investing in AI literacy, critical thinking, and adaptability, which are the focus of AI talent in Chapter 11. The Microsoft report indicates that 47% of leaders list upskilling as a top strategy.
- Further hurdles are ensuring ethical governance and data security to build trust, as detailed in Chapter 8.
- Integrating disparate technology systems to ensure AI access and data use, leveraging APIs and automation, as covered in Chapter 13.
- Managing the investment by justifying and managing resources through clear use cases and ROI, which are discussed in various sections like Chapter 4 and Appendix C.

The Four Audiences of an AI Nexus Hotel

The AI Nexus Hotel operates in a complex information ecosystem. Understanding and optimizing for each of these four audiences is pivotal for success in an AI-driven hospitality future:

1. **Human Guests and Staff:**

    - *Who they are:* This is your traditional and most important audience. The individuals physically present at your property, including guests seeking memorable experiences and the staff dedicated to creating them.

- *Why they matter in an AI Nexus Hotel:* While AI transforms operations, the core mission of hospitality remains centered on the human experience. AI aims to enhance guests' stays through seamless service, personalization, and quicker responses, freeing up staff from toil to deliver more joy through genuine interaction and empathetic service. AI tools should empower staff, make their work more fulfilling, and reduce burnout.

2. **AI Agents Working for Humans (External AI Assistants):**

   - *Who they are*: These are AI systems or assistants that act on behalf of external human users, such as a guest's personal AI travel agent tasked with booking accommodations or a corporate AI assistant researching venues for a meeting. These agents will increasingly interact directly with your hotel's booking systems or information portals.
   - *Why they matter in an AI Nexus Hotel:* As AI assistants become more prevalent in planning and booking travel, your hotel needs to be easily discoverable and bookable by these automated entities. If an external AI agent cannot easily find, understand, or interact with your hotel's offerings, you risk losing potential bookings.
   - *Optimizing for them:* This requires ensuring your hotel's distribution channels, website, and booking engines are equipped with clear, well-structured data and robust APIs (application programming interfaces). Information about room types, availability, pricing, amenities, and policies must be transparent

and easily parsable by these external AI agents to facilitate seamless, automated bookings and information gathering.

3. **Your AI Systems (Internal Digital Labor):**

    - *Who they are:* This audience comprises your hotel's internal AI tools, including the custom GPTs (Chapter 12), intelligent workflow automation (Chapter 13), and AI agents (Chapter 14) that you deploy. These systems function as your digital labor.
    - *Why they matter in an AI Nexus Hotel:* Your internal AI's efficiency, accuracy, and effectiveness depend on the quality of the data and instructions they receive. They are a critical component of your operational backbone.
    - *Optimizing for them:* This demands high-quality, clean, accurate, and well-structured internal data for training and operation. SOPs and knowledge bases used by these AI systems must be current and unambiguous. Ensuring seamless integration and communication between your internal AI tools is vital for creating efficient, end-to-end automated processes.

4. **External AI Systems:**

    - *Who they are:* This broad category includes AI-powered search engine crawlers (like those used by Google or Bing for their AI-driven search summaries), chatbots like ChatGPT and Gemini, AI models used by online travel agencies (OTAs) and travel aggregators to analyze and present hotel options, and other AI systems that learn from publicly available data about your hotel.

- *Why they matter in an AI Nexus Hotel:* How these external AI systems perceive and interpret your hotel significantly impacts your online visibility, reputation, and how you are presented to potential guests in AI-generated summaries or recommendations. Inaccurate or poorly structured public data can lead to misrepresentation.
- *Optimizing for them*: This involves proactively managing your hotel's publicly available structured data, for instance, by using schema.org markups on your website to help AI understand your content. Ensuring your website is easily crawlable, that information about your services and amenities is clear and consistent across all platforms, and monitoring how AI tools summarize or describe your property are key strategies.

By recognizing and strategically catering to these four distinct audiences, the AI Nexus Hotel can maximize its operational efficiency, enhance guest and staff experiences, improve market visibility, and ensure its digital labor performs optimally.

The AI-Integrated Future

The AI Nexus Hotel is about marshaling AI to craft experiences that are more deeply personal, memorable, and genuinely human than ever before. Those who will thrive are the visionaries who understand that AI's role is to amplify their unique expression of hospitality, ensuring the human touch, elevated beyond toil, remains the heart of the experience. AI's most powerful gift will be its ability to enable our human interpersonal capabilities. This revolution is about reimagining and reconstructing work itself with AI as a core collaborator. Therefore, the critical question is, what conviction and speed will we choose to architect that future?

Key Takeaways:

- The AI Nexus Hotel represents a future hospitality model deeply integrated with AI, mirroring Microsoft's Frontier Firm, where AI acts as intelligent connective tissue across operations, data, guest experiences, and human talent. This is driven by intelligence on tap and necessitates new organizational structures like human-agent teams and work charts.
- Progressing toward an AI Nexus Hotel involves moving through AI ascension levels, which align with broader industry phases of AI adoption, starting with AI as an assistant, then as a digital colleague, and ultimately to humans setting direction for agent-run processes.
- In the AI Nexus Hotel, the human role is elevated to focus on empathy, creative problem-solving, ethical governance, and strategic leadership, with every employee potentially becoming an agent boss. A humanity-first, AI-powered ethos guides this.
- Successfully becoming an AI Nexus Hotel requires overcoming hurdles, such as fostering leadership vision, managing cultural change, developing new skills (as highlighted by the Microsoft report), ensuring robust ethical governance, integrating technologies, managing investments, and understanding the needs of four distinct audiences, including AI systems themselves.

Bringing This into Focus:

- **Assess Your Hotel's Current AI Ascension Level This Month**: Using the AI ascension levels described in this chapter and considering Microsoft's three phases of AI integration, conduct an honest assessment with your team to determine where your hotel or department currently stands. Identify one key action you can take in the next three months to realistically move toward the next level.

- **Envision One AI Nexus Hotel Characteristic to Pilot:** From the descriptions of the AI Nexus Hotel, such as the work chart concept, the agent boss role, or leveraging intelligence on tap, pick one characteristic that most stimulates you or seems most achievable in the near term. Brainstorm one small, practical step your team could take next quarter to begin moving toward realizing that characteristic in your specific context.
- **Identify Your Biggest Adoption Hurdle and a Corresponding "Path Forward" Step:** Review the common hurdles to becoming an AI Nexus Hotel (leadership, culture, skills, ethics, integration, investment). Identify which one represents your biggest current barrier. Then, consider one of Microsoft's "Path Forward" steps (hire digital employees, set human-agent ratio, scale fast) and discuss how it could specifically help address that hurdle in your organization.

Leadership Action Agenda:

- **Conduct an AI Nexus Hotel Visioning Workshop with Leadership:** Convene your senior leadership team to discuss Microsoft's Frontier Firm concept and its application as the AI Nexus Hotel for your organization. Use the AI ascension levels and Microsoft's phases as a framework for self-assessment. Collaboratively define what becoming AI-native or AI-integrated means for your hotel's brand, guest experience, and operational model over the next one to three years, considering the urgency highlighted by current AI adoption trends.
- **Design a Human-Agent Team and Work Chart Pilot Project:** Select one key department or a significant cross-functional process within your hotel. Task a working group with designing a pilot AI-augmented team structure for that area, incorporating the concept of digital labor and considering the work chart model. This design should

outline potential new role responsibilities for humans and AI, including agent boss considerations, revised workflows, and necessary skill shifts. Focus on how AI could create more time for strategic work and enhance overall team effectiveness.

- **Develop and Communicate a Humanity-First, AI-Powered Charter:** Draft a clear set of guiding principles that articulate your hotel's commitment to using AI while prioritizing human well-being, ethical integrity, and genuine hospitality. This charter should align with the ethos of AI-operated and human-led systems found in the Frontier Firm and embodied by the AI Nexus Hotel. Share this charter widely within your organization to set the cultural tone for your AI ascension and reinforce the message that technology serves to amplify human capabilities.

EPILOGUE

AGI ON THE HORIZON

The technology we have explored throughout this playbook is already significantly changing hospitality. Yet, off in the distance, closer than many believe, lies a profound shift that will dwarf everything we have discussed: artificial general intelligence, or AGI. While experts debate precise timelines, with predictions ranging from a few short years to several decades, the accelerated pace of AI development is undeniable. The race toward this technology involves more than just another Silicon Valley competition; it is a high-stakes sprint that is reshaping the future of computing itself. OpenAI explicitly states its mission is "to ensure that artificial general intelligence benefits all of humanity." Google DeepMind has invested billions in pursuing this goal. Anthropic was founded specifically to safely develop AGI. Even traditional tech giants like Microsoft, Meta, and Apple are pouring unprecedented resources into this pursuit. Why? Because whoever achieves AGI first will control an entirely new computing paradigm.

To understand AGI, we need to clarify what it is not. Today's AI tools, from ChatGPT to Gemini to the custom GPTs we explored in Chapter 12, are all examples of narrow AI. These systems excel at specific tasks they were designed for but fail completely outside their fields. A language model can write brilliant marketing copy but cannot recognize a guest's face. An image generator can create stunning visuals but cannot book a reservation. A revenue management system can optimize pricing but cannot have a detailed conversation with an angry guest. This is the fundamental limitation of narrow AI: Each system is specialized and extraordinarily capable within strict boundaries but unable to transfer knowledge or skills between fields.

AGI breaks through this barrier. It could understand, learn, and apply its intelligence across any field, similar to human cognitive flexibility. While narrow AI needs extensive retraining for new problems, AGI could learn continuously, reason through novel situations, and apply concepts across fields without specialized reprogramming. For hospitality, imagine a system that could simultaneously analyze your financial performance, craft personalized marketing campaigns, optimize your revenue strategy, manage your property's energy consumption, and handle guest communications, all with human-level competence and at machine speed and scale. A front desk agent could ask this system, "What is the best approach for the Smith family arriving today?" and receive an answer that integrates their reservation history, preferences, current property occupancy, weather, and even social media sentiment, all contextualized into actionable recommendations. Or envision an AGI for strategic hotel management, capable of analyzing global travel trends, local event impacts, competitor strategies, and real-time operational data to propose novel, highly adaptive business models or guest experience concepts for human leadership to consider and approve.

The promise is extraordinary. Hotels could achieve unprecedented efficiency while delivering hyper-personalized experiences at scale. An AGI system could understand guests, perhaps more comprehensively than they understand themselves, anticipating needs before they arise. Property management could become largely autonomous, with human oversight focused on strategic decisions and exceptional circumstances.

Still, legitimate concerns shadow these possibilities. What happens to millions of hospitality workers whose jobs could be fundamentally altered or eliminated? How do we maintain the essential human elements that define hospitality—empathy, cultural distinction, genuine warmth—when so much is automated? Who controls these systems, and what happens when AGI's capabilities expand beyond its creators' intentions? What ethical frameworks guide AGI systems in handling sensitive guest data or making decisions that impact people's lives? These questions are more than philosophical; they are practical business considerations as AGI draws closer.

Despite these profound challenges, I remain deeply optimistic about hospitality's future in the AGI age. Throughout history, technological revolutions have consistently significantly changed jobs rather than eliminating them. The ATM was predicted to eliminate bank tellers; instead, it changed their role to relationship managers and financial advisors. Accounting software failed to kill the accounting profession; it elevated accountants to strategic business partners. AGI will elevate, not replace, hospitality professionals. This elevation occurs as AGI shoulders the immense burden of complex data processing and routine operational management, freeing human talent to focus on what they do best: sophisticated strategic oversight, creative innovation in guest experience design, and the cultivation of deeply empathetic human connections. In an industry as deeply rooted in genuine human interaction and detailed service as hospitality, AGI's role as an amplifier of human capability, rather than a replacement, is perhaps even more pronounced, promising a future where technology handles complexity so humans can perfect the art of connection. Freed from routine operations and armed with unprecedented insights, tomorrow's hoteliers will become experience architects, emotional intelligence experts, and masters of the authentic human connection that no machine, however intelligent, can fully replicate.

The properties that succeed will be the ones that thoughtfully integrate these systems into a human-centered vision of hospitality, using AGI to handle the predictable so their teams can focus on creating the remarkable. This is not science fiction. Thousands of researchers are now assembling the building blocks of AGI. Companies are making monthly breakthroughs. The capabilities we discuss as conceptual today could arrive with stunning speed tomorrow. That is why the principles explored throughout this playbook, from AI literacy to ethical governance, from workflow automation to the human in the loop, are the essential foundation for thriving in the AGI-powered future that awaits beyond the horizon.

The journey we have taken together is far from ending; it is just beginning. While none of us can perfectly predict where this technology will lead, I believe this with certainty: The future of hospitality belongs to those who embrace these tools as amplifiers of our unique capacity to create moments of joy, comfort, and meaning for every guest we serve. The AI revolution is about technology finally giving us the space to be more deeply authentically human than ever before.

APPENDICES

APPENDIX A

USE CASE IDENTIFICATION WORKSHEET

This Use Case Identification Worksheet is your guide, designed to lead your team through a process for pinpointing AI use cases within your hotel. It operationalizes principles from this playbook for understanding where AI can make an impact by analyzing work processes and identifying areas of toil (as introduced in the Introduction and explored with the joy-toil spectrum in Chapter 10).

To leverage this worksheet for impact on your AI ascension, approach it with these principles:

- The worksheet yields insights when completed by teams or department representatives. Different perspectives enrich the process.
- Do not expect perfection in one session. Revisit and refine as your understanding of AI and operations deepens.

- The more candid your team is in assessing pain points and toil, the more valuable your identified AI use cases will be.
- The objective is to move toward AI pilot projects that deliver results.

Here is a look at what each part of the worksheet will guide you to accomplish:

Part 1: Defining Your Focus Area (Department/Process)

This section helps you define the scope of your analysis for targeted effort. You will:

- Identify the department (e.g., front office, sales) or cross-departmental process (e.g., guest complaint resolution) to analyze.
- List the key roles involved.
- Clarify the primary goals of this department or process.
- Pinpoint the current challenges or pain points within this focus area. This sets the stage for identifying where AI can provide relief and value.

Part 2: Breaking Down the Work From Roles/Processes to Granular Tasks

This is where analysis begins. Referencing principles of work analysis and toil identification (Chapter 10 and Introduction), you will:

- List the main tasks or core responsibilities for each key role or major step in the process from Part 1.
- Then, deconstruct selected major tasks into specific sub-tasks needed for completion.

- For each sub-task, your team will:
    - Estimate current manual effort (time/frequency).
    - Assign a toil score (one for low toil/joyful work to five for high toil/draining work).
    - Assess its AI exposure (low, medium, or high), indicating how AI could assist or automate.
- If a sub-task is time-consuming or scores high on toil, break it down into sub-sub-tasks for more granular analysis. This deconstruction is key for pinpointing where AI interventions yield benefits.

Part 3: Task Analysis—Identifying AI Opportunities

With tasks deconstructed, this section guides you to brainstorm AI solutions. You will:

- Review tasks from Part 2, noting those with high toil scores, manual effort, and AI exposure.
- For each promising task or sub-task, brainstorm:
    - How AI could help (e.g., automate data entry, draft responses, summarize information, generate content variations)
    - Which AI functionality type is most relevant (e.g., creative, predictive, analytical, or automated)
    - What AI tool or platform might be suitable, if known (or general AI chatbot if unsure)
    - The desired outcome of using AI (e.g., reduce task time by X%, improve response quality, increase consistency, free Y hours for strategic work)

Part 4: Prioritizing AI Use Cases and Defining Next Steps

The final part helps you move from ideas to an action plan. You will:

- Select the top one to three AI use cases from Part 3 that offer the best combination of:
  - High Impact (time savings, toil, quality improvement, strategic value)
  - Feasibility (can be implemented with current or planned resources)
  - Team Enthusiasm (your team is motivated by the potential improvement)
- For each priority use case, you will collaboratively:
  - Articulate why it was selected
  - Outline the next two to three steps to explore it further (e.g., research AI tools, draft a sample prompt, mock up a workflow, define pilot success metrics)
  - Assign responsible person(s) and set a timeline

This worksheet provides a structured starting point for your AI initiatives.

To download the Use Case Identification Worksheet and begin charting your hotel's path to an AI-augmented future, visit http://vivanderadvisors.com/the-ai-literacy-playbook.

APPENDIX B

EXAMPLE THEMES FOR AI USAGE POLICIES

Note to the hotelier: The following themes are distilled from common AI usage guidelines often provided by AI model developers and external organizations. They are presented here as a reference for your AI governance committee when they undertake the task of drafting your hotel's specific AI governance policy, as guided by Chapter 8 of this playbook. These general themes are not a substitute for your comprehensive, hotel-specific policy but may help identify common areas that external AI usage policies often address. Your hotel's policy should reflect your unique brand values, operational needs, guest expectations, and legal/ethical obligations.

Common Themes Found in External AI Usage Guidelines

- **Legal and Ethical Adherence:**
    - A commitment to ensuring AI use complies with all applicable laws and regulations
    - Prohibitions against using AI for illegal activities, including the development or distribution of illegal goods/services or activities that could exploit or harm individuals, especially vulnerable populations like children
    - Stipulations for adhering to regulations when AI is used in regulated activities

- **Responsible Content Generation and Use:**
    - Restrictions on generating content that promotes or incites hate speech, harassment, defamation, violence, or discrimination against individuals or protected groups
    - Prohibitions on creating or distributing content intended to defraud, scam, mislead, or impersonate others
    - Guidelines against generating unsolicited or inappropriate sexually explicit content or gratuitous gore, except where it might be contextually appropriate for specific, approved purposes such as scientific, historical, or news discussions with proper warnings
    - Cautions against using AI to generate content that contributes to extremist agendas or promotes violence

- **Respect for Privacy and Intellectual Property:**
    - A strong emphasis on protecting individuals' privacy by not using AI to request, generate, or disseminate private, personal, or sensitive information without explicit consent or legitimate purpose
    - A commitment to respecting the rights of creators and intellectual property holders, avoiding the use of AI to knowingly infringe on copyrights, trademarks, or other IP rights

- **Safe and Appropriate User Interaction and System Behavior:**
    - Guidelines for AI systems to interact respectfully and helpfully, assuming the best intentions from users where appropriate
    - Encouraging AI systems to ask clarifying questions when user intent is unclear and to express uncertainty or limitations where they exist
    - Directives for AI to maintain objectivity, avoid promoting personal agendas, and not be sycophantic
    - Prohibitions against using AI for targeted manipulation of political views or to spread misinformation that could cause societal harm

- **Technical and Operational Soundness:**
    - Ensuring AI systems are used efficiently and that outputs are generated in appropriate and usable formats, for instance, using Markdown with LaTeX extensions where specified for technical content
    - Guidelines for adapting AI responses respectfully and appropriately for different interaction

modalities, such as audio or video conversations, including gracefully handling interruptions

Reviewing these common external themes can provide a useful starting point for discussion as your AI governance committee develops a policy perfectly suited to your hotel's unique operational and ethical landscape.

APPENDIX C

AI IMPLEMENTATION ROADMAP

This appendix offers a phased approach to guide your hotel's AI implementation journey, building upon the insights from previous chapters. It emphasizes building a strong foundation, starting with practical applications and then strategically scaling. It is a flexible guide to help you manage this significant process.

To download the AI Implementation Roadmap, visit http://vivanderadvisors.com/the-ai-literacy-playbook/.

Key Principles for Your AI Implementation Journey

- *Human-Centric Approach:* AI should augment human capabilities, reduce toil, and enhance joy instead of just replacing tasks.
- *Strategic Alignment:* Ensure every AI initiative aligns with your hotel's overarching business goals and guest experience vision.

- *Iterative Progress:* Start small, demonstrate value, learn, and scale. Perfection is not the initial goal; practical progress is.
- *Ethical Governance:* Embed responsible AI principles (Chapter 8) from the outset.
- Continuous Learning: Foster a culture where your team continuously learns about and adapts to developing AI capabilities.

Phase 1: Foundation and Awareness (Months 1–3) "Preparing the Ground"

Goal: Build foundational AI literacy, establish initial governance, and identify early quick win opportunities.

Key Activities:

- Leadership Alignment and AI Vision Kick-off: Conduct workshops with the senior leadership team to discuss AI's potential impact, both positive and negative, on the hotel. Define a high-level vision for how AI will support the hotel's strategic objectives and brand values.
- Establish AI Governance Committee and Draft Initial Policy: Form your cross-functional AI governance committee (Chapter 8). Task them with drafting a V1.0 AI governance policy focusing on data privacy, ethical use, and approved tools.
- Roll Out Foundational AI Literacy Training: Implement AI essentials training for all relevant staff, focusing on basic AI concepts, the new remote associate mindset, and core prompting skills.
- Identify and Train Initial AI Champions: Select enthusiastic individuals from key departments to become AI champions. Provide them with slightly more advanced training.

- Conduct Departmental Use Case Identification Workshops: Using the worksheet in Appendix A and concepts for analyzing work, facilitate workshops in one to two pilot departments to identify high-toil, high-impact AI use cases.
- Select One or Two Quick Win Pilot Projects: Choose highly feasible projects that can demonstrate AI's value quickly, for example, using AI to draft standard email responses for one department or creating a simple custom GPT for internal FAQs.

Expected Outcomes: Basic AI awareness across the organization, initial AI governance framework in place, a pool of identified potential AI use cases, one to two small pilot projects initiated, and increased team curiosity and reduced apprehension about AI.

Phase 2: Pilot Implementation and Active Learning (Months 4–6) "Testing the Waters"

Goal: Implement initial pilot projects, gather learnings, refine processes, and build practical AI application skills.

Key Activities:

- Execute and Monitor Pilot Projects: Deploy the chosen quick-win AI solutions in the pilot departments. Closely monitor their usage, effectiveness, and impact on toil time.
- Gather Feedback and Iterate: Actively solicit feedback from users of the pilot AI tools. Use this feedback to iterate on prompts, custom GPT instructions, or workflow designs.
- Develop Department-Specific AI Training: Driven by pilot learnings, create more customized AI training modules for specific departmental needs and use cases, such as AI for Sales Prospecting or AI for Marketing Content.

- Refine AI Governance Policy and Processes: The AI governance committee reviews early learnings and refines the AI policy and approval processes.
- Begin Building Shared Prompt Libraries: Encourage pilot teams to document successful prompts and start creating departmental shared prompt libraries.
- Explore Initial Workflow Automation Opportunities: Identify one to two simple, high-impact cross-system workflows that could be automated.

Expected Outcomes: Successful completion of one to two pilot AI projects with measurable benefits, increased practical AI skills within pilot departments, refined AI governance policies informed by real-world use, first shared prompt libraries established, growing confidence and enthusiasm for AI across wider teams, and initial understanding of workflow automation potential.

Phase 3: Scaling Successful Pilots and Deeper Integration (Months 9–12) "Expanding the Footprint"

Goal: Expand successful AI solutions to other departments, deepen AI integration into core processes, and mature automation capabilities.

Key Activities:

- Scale Proven AI Solutions: Take learnings from successful pilots and strategically roll out those AI tools/custom GPTs to other relevant departments.
- Invest in More Advanced AI Tools and Platforms: According to the needs identified, consider investing in more sophisticated AI platforms, such as enterprise AI solutions or advanced automation tools, if justified by ROI.
- Deepen Workflow Automation: Implement more complex intelligent workflow automation to streamline key operational processes and connect more systems.

- Advanced Custom GPT Development: Build more sophisticated custom GPTs with more extensive knowledge bases and potential actions for system integration.
- Ongoing AI Skill Development and Specialization: Offer advanced training for AI champions and opportunities for team members to specialize in areas like AI ethics, prompt engineering, or AI workflow design.
- Regularly Review and Update AI Strategy and Governance: The AI governance committee continues to refine policies and ensure the overall AI strategy aligns with developing business needs and technological advancements.

Expected Outcomes: Wider adoption of proven AI solutions across the hotel; increased operational efficiency and measurable time savings; more sophisticated AI applications, like custom GPTs with actions or multi-step automation; a growing pool of AI-skilled employees and active AI champions; AI ethics and governance deeply embedded in operations; and a clear roadmap for further AI integration and innovation.

Phase 4: Optimization, Innovation, and Continuous Development (Months 12+) "Becoming AI-Native"

Goal: Foster a culture of continuous AI-driven improvement, explore advanced AI applications like AI agents, and solidify the hotel's position as an AI Nexus Hotel.

Key Activities:

- Optimize Existing AI Implementations: Continuously monitor and refine AI tools, workflows, and custom GPTs for peak performance and user satisfaction.
- Explore Advanced AI Agents: Begin to explore the potential of AI agents for more complex, goal-oriented tasks, starting with well-defined, low-risk pilot projects.

- Focus on AI-Driven Strategic Insights: Leverage AI for deeper data analysis to inform strategic decision-making across revenue, marketing, operations, and guest experience.
- Embed AI in Innovation Processes: Integrate AI tools into brainstorming, product development, and service design processes to accelerate innovation.
- Foster a Culture of Continuous Learning and Adaptation: Ensure AI skills development is an ongoing priority, with mechanisms for sharing knowledge and adapting to new AI breakthroughs.
- Proactive Ethical Oversight and Future-Readiness: The AI governance committee focuses on anticipating future ethical challenges and opportunities presented by advancing AI, ensuring the hotel remains a leader in responsible AI use.

Expected Outcomes: A deeply AI-integrated hotel operation, an AI Nexus Hotel; AI is a core component of strategic decision-making and innovation; a highly AI-literate workforce, comfortable collaborating with AI; ongoing optimization and development of AI applications; and recognition as an industry leader in leveraging AI for exceptional hospitality.

This phased roadmap provides a strategic framework. Your hotel's specific timeline and priorities will vary. The key is to start, learn, adapt, and always keep the human element at the center of your AI journey.

Cross-Cutting Themes for All Phases

Throughout all phases of your AI implementation, several cross-cutting themes must remain constant priorities:

- *Ethical Governance and Human Oversight:* Continuously apply your AI governance framework, ensuring all AI use is ethical, responsible, safe, and always includes a human in the loop for critical decisions.

- *Data Security and Privacy (Chapter 8):* Maintain rigorous data security and privacy protocols for any data AI systems use or generate.
- *Change Management and Communication:* Proactively manage the human side of AI adoption. Communicate transparently with your teams about AI initiatives, involve them in the process, address concerns, and highlight how AI will augment their roles and reduce toil.
- *Continuous Learning and Adaptation:* The AI situation is rapidly developing. Foster a culture of continuous learning, experimentation, and adaptation to ensure your hotel stays ahead of the curve.
- *Collaboration and Knowledge Sharing:* Encourage collaboration between departments and the sharing of AI best practices, successful prompts, and innovative use cases to accelerate learning across the organization.

Measuring Impact and ROI: Key Performance Indicators for AI Success

From the outset, define how you will measure your AI initiatives' success and return on investment. While traditional ROI can be challenging to pinpoint for a technology that's rapidly becoming a cost of doing business, focusing on usage and the value derived from amplified human capabilities offers a clearer picture. As you progress through your AI implementation, track KPIs across several key areas to demonstrate value. Remember that if people are using these tools consistently, it's a strong signal they are finding them valuable and are amplifying their abilities. Consider the following KPIs:

I. Usage and Adoption KPIs (Foundation of Value)

These are leading indicators. Consistent usage suggests perceived value and is the first step to unlocking other benefits.

- **Daily/Weekly Active AI Users (DAU/WAU):**
  - *What it measures*: The number or percentage of targeted staff actively using the AI tools daily or weekly.
  - *Why it matters*: High DAU/WAU indicates the tool is becoming integral to daily workflows rather than a novelty. If people are using it, they are likely deriving benefits.
  - *Target*: Set a realistic adoption target, for example, 70% of licensed users active weekly after three months.

- **AI Task Completion Rate/Volume:**
  - *What it measures*: If your AI tools allow, track the number of specific tasks completed using AI, such as emails drafted, reports summarized, queries run, or content generated.
  - *Why it matters*: Shows the volume of work being augmented or handled by AI, indicating practical application.
  - *Target*: Increase in volume of AI-assisted tasks over time.

- **Feature Adoption Rate:**
  - *What it measures*: Which specific AI features or custom GPTs are most frequently used.
  - *Why it matters*: It helps understand which functionalities provide the most value and where further training or promotion might be needed.
  - *Target*: Identify key value-driving features and aim for high adoption in relevant departments.

II. Efficiency and Productivity KPIs (Leveraging Saved Time)

This is where benefits, like the one hour saved daily, start translating into broader impacts.

- **Perceived Time Savings and Reallocation:**
    - *What it measures*: Staff-reported estimates of time saved per day/week using AI via regular, simple surveys *and* how they are reallocating that saved time, for instance, to more strategic work, higher quality output, or direct guest interaction.
    - *Why it matters*: Quantifies and links a primary benefit to value generation.
    - *Target*: Achieve an average self-reported time savings of X hours/week per user, with a significant portion of that time reallocated to value-added activities.

- **Reduction in Toil Task Time:**
    - *What it measures*: Decrease time spent on specific, highly repetitive, low-value tasks identified during your job deconstruction and joy-toil audits (concepts discussed in Chapter 10).
    - *Why it matters*: Directly shows AI alleviating burdensome work, freeing up human capacity for more engaging tasks.
    - *Target*: X% reduction in time spent on top three toil tasks per department within six months.

- **Process Cycle Time Reduction:**
    - *What it measures*: The average time it takes to complete specific end-to-end processes where AI

has been implemented, for example, generating a group sales proposal, resolving a guest complaint, or onboarding a new employee.
  - *Why it matters*: Faster cycles often mean better guest service, quicker revenue generation, or improved operational agility.
  - *Target*: Reduce cycle time for X key processes by Y% within a year.

III. Quality and Effectiveness KPIs (Amplifying Abilities)

If AI is amplifying abilities, the quality of work should improve.

- **First-Pass Quality Improvement:**
  - *What it measures*: Reduction in errors, rework, or necessary revisions for tasks where AI provides a first draft or assistance, such as fewer typos in AI-drafted emails or more complete initial reports.
  - *Why it matters*: AI helps produce higher-quality work from the start, saving further human editing time and improving overall output.
  - *Target*: Measurable reduction in common errors or revisions for specific AI-assisted tasks.

- **Consistency Metrics:**
  - *What it measures*: Improved adherence to brand guidelines in communications if using a brand guardian custom GPT, consistency in responses to standard guest inquiries, or uniform application of SOPs where AI assists.

  - *Why it matters*: AI can help ensure a more consistent guest experience and brand representation.
  - *Target*: Audit a sample of AI-assisted outputs monthly and track adherence to standards.

IV. Staff Empowerment and Engagement KPIs (The Joy Factor)

Usage should ideally correlate with staff feeling more capable and less burdened.

- **Employee Joy vs. Toil Score Shift:**
  - *What it measures*: Changes in staff perception of their work, moving from toil to joy for tasks now augmented by AI; revisit the audit concepts from Chapter 10.
  - *Why it matters*: This directly reflects AI's impact on the quality of work-life and engagement.
  - *Target*: Increase in joy scores and decrease in toil scores for AI-impacted roles within twelve months.
- **AI-Powered Innovation Submissions:**
  - *What it measures*: The number of new ideas, process improvements, or AI use cases suggested by staff members.
  - *Why it matters*: It shows that the team is thinking creatively about how it can further benefit the hotel, a sign of true engagement and an AI mindset.
  - *Target*: Implement a system for collecting AI ideas and aim for X number of actionable suggestions per quarter.

V. Strategic and Operational Impact KPIs (Moving the Needle)

- **Guest Satisfaction Uplift in AI-Impacted Areas:**
    - *What it measures*: Improvements in specific guest satisfaction metrics, such as survey scores for speed of service, staff helpfulness, or personalized attention, where AI supports the interaction.
    - *Why it matters*: Connects AI implementation to the core business of guest happiness.
    - *Target*: X% improvement in guest satisfaction for AI-supported touchpoints.
- **Strategic Project Advancement:**
    - *What it measures*: Progress or successful completion of strategic projects previously stalled or slow-moving due to lack of team bandwidth, bandwidth now potentially freed up by AI.
    - *Why it matters*: Demonstrates that saved time is being converted into strategic progress.
    - *Target*: Identify one to two strategic projects to be accelerated by AI-enabled teams and track their progress.

Important Considerations for Tracking KPIs

- *Baseline Measurement:* Where possible, establish a baseline measurement for these KPIs before widespread AI implementation. This will allow you to demonstrate the impact and improvements over time clearly.
- *Qualitative Feedback:* Supplement these quantitative KPIs with qualitative feedback. Conduct regular, informal check-ins with staff, include AI-related questions in

employee engagement surveys, and gather anecdotal evidence of AI's impact on their work and guest interactions. This provides the story and context behind the numbers.

- *Iterative Approach to KPIs:* Recognize that your most relevant KPIs might evolve as your hotel's AI usage matures and AI technology develops. Review and adjust your KPIs periodically to ensure they align with your AI strategic goals.

APPENDIX D

GLOSSARY OF AI TERMS FOR HOTELIERS

This glossary defines key AI terms used throughout this playbook, helping you confidently speak the language of AI.

- AI (Artificial Intelligence): The capability of computer systems to perform tasks that typically require human intelligence, such as learning, problem-solving, and decision-making.
- AI Agent: An AI system capable of perceiving its environment, reasoning, planning, and taking autonomous actions to achieve specific goals (see Chapter 14 for AI agents).
- AI Ascension: A hotel's strategic journey to integrate AI deeply into its operations, progressing through defined levels of AI literacy, adoption, and, ultimately, AI-native status (see Chapter 15 for AI Nexus Hotel).

- AI Champions: Team members within various hotel departments trained to be internal advocates, peer mentors, and first-line support for AI tools and responsible AI use (see Chapter 8 for AI governance).
- AI Exposure Levels: A framework for assessing how significantly a specific task or sub-task will likely be impacted or augmented by current and emerging AI capabilities (refer to concepts discussed in Chapter 10 on work analysis for AI opportunities).
- AI Governance: The comprehensive framework of policies, processes, roles, and ethical guidelines that direct and oversee the responsible, ethical, and safe use of AI within an organization (see Chapter 8 for AI governance).
- AI Governance Committee: A cross-functional team responsible for developing, implementing, and overseeing the hotel's AI governance policies, ensuring ethical AI use and strategic alignment (see Chapter 8 for AI governance).
- AI Literacy: A foundational understanding of what AI is, how it works, its capabilities and limitations, and how to interact with it effectively, responsibly, and ethically (see Chapter 2 for AI literacy and mindset).
- AI Mindset: An adaptive approach to thinking strategically, critically, and collaboratively with AI, viewing it as a partner to augment human capabilities and enhance work (see Chapter 2 for AI literacy and mindset).
- AI Nexus Hotel: This playbook's term for a hotel that has deeply integrated AI into its operational fabric, becoming an AI-native organization where AI acts as intelligent connective tissue, amplifying human talent and significantly changing guest experiences (see Chapter 15 for AI Nexus Hotel).
- Algorithm: A set of rules or instructions given to an AI system or computer to help it learn, make decisions, or solve problems.

- Analytical AI: AI that sifts through existing data, such as guest reviews or sales figures, to find hidden patterns, themes, correlations, or anomalies, helping to understand what happened.
- API (Application Programming Interface): A set of rules and protocols that allows different software applications to communicate and exchange data with each other, crucial for enabling AI tools to interact with hotel systems like PMS or CRM (see Chapter 13 for intelligent workflow automation).
- Artificial General Intelligence (AGI): A conceptual future form of AI with cognitive abilities comparable to, or surpassing, human intelligence across a broad range of intellectual fields (see Epilogue).
- Automation (Intelligent Workflow Automation): Connecting different software systems and AI tools so they can work together proactively and automatically, triggered by events, to perform multi-step tasks or processes with reduced human intervention (see Chapter 13 for intelligent workflow automation).
- Bias (in AI): Systematic errors or unfair skews in AI outputs resulting from biases in the AI's training data or algorithm design, potentially leading to discriminatory or inequitable outcomes (see Chapter 8 for AI governance).
- Chatbot: An AI program designed to simulate human conversation through text or voice interactions, often used for customer service, information retrieval, or task assistance.
- Cognitive Offloading: The tendency to over-rely on external tools like AI to perform thinking tasks, potentially diminishing one's critical thinking abilities or memory retention over time if lacking balance with active mental engagement (see Chapter 9 for critical thinking).
- Context Window: The amount of information, such as text or data, an AI model can "remember" or consider at one time during a specific conversation or task. Larger context windows

allow AI to handle more complex inputs and maintain coherence over longer interactions (see Chapter 4 for AI toolkit).

- Counter Prompting: An advanced prompting technique where you ask an AI to argue against its suggestion or critique your proposed idea to uncover weaknesses or consider alternative perspectives (see Chapter 9 for critical thinking).
- Critical Thinking (in AI Context): The human ability to analyze and evaluate AI-generated outputs for accuracy, relevance, bias, and logical soundness before accepting or acting upon them (see Chapter 9 for critical thinking).
- Custom GPT: A user-created, specialized version of a general AI model like ChatGPT that has been customized with specific instructions, knowledge, and capabilities to perform particular tasks or adopt a defined persona for a specific context, such as a hotel (see Chapter 12 for custom GPTs).
- Data Privacy: Protecting the sensitive personal information of guests or employees from unauthorized access, use, or disclosure, especially when using AI systems (see Chapter 8 for AI governance).
- Deep Learning: A subset of machine learning that uses artificial neural networks with many layers (deep networks) to learn complex patterns from large amounts of data.
- Digital Labor: AI systems, particularly AI agents, that are formally integrated into organizational workflows to perform tasks, augmenting human labor and contributing to operational output (see Chapter 15 for AI Nexus Hotel).
- Ethical AI: Designing, developing, and deploying AI systems in a way that aligns with human values, promotes fairness, avoids harm, and ensures accountability (see Chapter 8 for AI governance).
- Few-Shot Prompting: An AI prompting technique where you provide the AI with one or more examples (shots) of the desired output format or style within the prompt itself to guide it more effectively (see Chapter 5 for prompting).

- Frontier Firm: Microsoft's term for an entirely new kind of organization structured around on-demand intelligence and powered by hybrid teams of humans and AI agents, representing the next development of business in the age of AI (see Chapter 15 for AI Nexus Hotel).
- Generative AI: A type of AI that can create new, original content, such as text, images, audio, or code, based on patterns and information learned from its training data (see Chapter 1 for "What is AI").
- Hallucination (AI): An instance where an AI model generates incorrect, nonsensical, or fabricated information and presents it as factual and confident output (see Chapter 8 for AI governance, Chapter 9 for critical thinking).
- Human-in-the-Loop (HITL): A crucial principle in AI implementation where humans actively oversee, review, validate, approve, or intervene in AI-driven processes or outputs, especially for critical or sensitive tasks (see Chapter 8 for AI governance, Chapter 13 for intelligent workflow automation).
- Intellectual Property (IP): Creations of the mind, such as inventions, literary and artistic works, designs, symbols, names, and images used in commerce. AI's role in content creation raises questions about IP ownership and infringement (see Chapter 8 for AI governance).
- Intelligence Resources: A conceptual future hotel department blending IT and HR capabilities. It is dedicated to enabling and optimizing the hotel's AI ecosystem and the collaboration between human talent and AI agents (see Chapter 15 for AI Nexus Hotel).
- Job Deconstruction: The process of breaking down a job role or process into its constituent major tasks, sub-tasks, and even sub-sub-tasks to identify specific areas where AI can assist or automate.
- Joy-Toil Spectrum: A framework for analyzing work tasks according to their emotional and cognitive impact,

distinguishing between joy (fulfilling, energizing, high-value work) and toil (draining, repetitive, low-value work), to guide AI implementation toward enhancing employee satisfaction (see Chapter 10).

- Knowledge Cutoff: The point in time beyond which an AI model's training data does not extend, meaning it generally lacks awareness of events or information that emerged after that date unless connected to live data sources (see Chapter 4 for AI toolkit).
- Large Language Model (LLM): An AI model trained on vast amounts of text data to understand, generate, and manipulate human language with impressive fluency, such as GPT-4, Claude 3, or Gemini.
- Machine Learning (ML): A subset of AI where systems learn from data, identify patterns, and make decisions with minimal human intervention, improving their performance over time through experience.
- Multimodality (AI): The ability of an AI system to process, understand, and generate information across multiple formats or modes, such as text, images, audio, and video (see Chapter 4 for AI toolkit, Chapter 7 for sights and sounds).
- Narrow AI (Artificial Narrow Intelligence, ANI): AI systems designed and trained for a specific task or a limited range of tasks, such as image recognition, language translation, or playing chess. All current AI applications are considered narrow AI.
- Natural Language Processing (NLP): A field of AI focused on enabling computers to understand, interpret, and generate human language in a meaningful and useful way.
- Neural Network (Artificial Neural Network, ANN): A computing system inspired by the structure and function of the human brain, consisting of interconnected nodes or neurons that process information in layers to learn patterns and make decisions.

- Predictive AI: AI that analyzes historical and current data to forecast future events, behaviors, or trends, such as predicting hotel occupancy, guest demand, or potential equipment failures.
- Prompt: A human's instruction, question, or input to an AI model to elicit a desired response or action (see Chapter 5 for prompting).
- Prompt Engineering: The art and science of crafting effective prompts to guide AI models to produce accurate, relevant, high-quality outputs (see Chapter 5 for prompting).
- Responsible AI: An approach to AI development and deployment that emphasizes accountability, transparency, fairness, reliability, and safety, ensuring AI systems operate as intended and serve beneficial purposes (see Chapter 8 for AI governance).
- Symbolic AI (Good Old-Fashioned AI, GOFAI): An early approach to AI that aimed to build intelligent systems by programming computers with explicit human knowledge, rules, and logical processes (see Chapter 1 for "What is AI?").
- Temperature (AI Setting): A parameter in many generative AI models that controls the randomness or creativity of the output. Lower temperatures lead to more predictable, common responses; higher temperatures encourage more novel or unexpected outputs.
- Text-to-Image Generation: An AI capability where a model generates a unique image driven by a textual description (prompt) provided by a user (see Chapter 7 for sights and sounds).
- Text-to-Video Generation: An AI capability where a model generates a video clip according to a textual description (prompt) (see Chapter 7 for sights and sounds).
- Text-to-Speech (TTS): An AI capability that converts written text into audible, spoken language (see Chapter 7 for sights and sounds).

- Token (AI Context): The fundamental unit of data, often a word, part of a word, or punctuation, that large language models process and generate when handling text (see Chapter 1 for "What is AI?" and Chapter 4 for AI toolkit).
- Training Data: The large dataset of information, such as text, images, or code, used to "teach" or train an AI model, from which it learns patterns, relationships, and knowledge (see Chapter 1 for "What is AI?", Chapter 4 for AI toolkit, and Chapter 8 for AI governance).
- Transformer Model: A type of deep learning architecture introduced in 2017 that uses an attention mechanism to weigh the importance of different parts of input data, enabling significant advances in natural language processing and forming the basis for most large language models (see Chapter 1 for "What is AI?").
- TUR Filter: A practical framework for daily, rapid assessment of AI-generated outputs, evaluating for truth/accuracy, usefulness, and contextual relevance before accepting or using the output (see Chapter 9 for critical thinking).
- Voice Cloning: An AI capability to synthesize speech that closely mimics a specific individual's voice, requiring ethical consideration and consent (see Chapter 7 for sights and sounds).
- Work Chart: A concept for the organizational structure of Frontier Firms/AI Nexus Hotels, emphasizing dynamic, outcome-driven human-agent teams assembled around goals or jobs to be done rather than traditional, fixed functional hierarchies (see Chapter 15 for AI Nexus Hotel).

This glossary gives you the language needed to confidently discuss, implement, and lead with AI in your hotel. Use it as a reference as you work through your AI ascension journey.

APPENDIX E

AI ASCENSION INITIATIVE PROPOSAL

This worksheet helps you outline a business case for a new AI initiative at your hotel. Its purpose is to clearly define a challenge or opportunity and propose how a specific AI solution can address it, contributing to your AI ascension.

To download the AI Ascension Initiative Proposal, visit http://vivanderadvisors.com/the-ai-literacy-playbook/.

1. Proposer and Focus Area:

    - Your Name: ____________________________
    - Your Role: ____________________________
    - Department/Process to Improve: ______________
    - Hotel/Location: ________________________

2. The Challenge/Opportunity:

   A. What is the specific business need, problem, or opportunity?

      (Briefly describe the current situation. What pain point needs solving, or what new value can be created?)

      ____

      ____

      ____

   B. Why is addressing this important right now?

      (Explain the impact. Consider effects on guest satisfaction, staff workload and toil vs. joy, operational efficiency, revenue, or strategic goals.)

      ____

      ____

      ____

   C. Proposed AI Solution and How It Works:

   A. Type of AI Solution Envisioned:

      (e.g., custom GPT, intelligent workflow automation, AI agent, or a specific AI functionality like generative, predictive, analytical, or repetitive AI).

      ____

      ____

      ____

B. Brief Description and Justification:

(How will this AI solution address the need? Why is it suitable? What will it do at a high level? Where is human-in-the-loop (HITL) oversight planned?)

____

____

____

4. Anticipated Benefits:

List the Top Two or Three Expected Benefits:

(Focus on measurable outcomes or clear improvements. Examples: X% time saved on [task], a significant reduction in [specific toil], increase in guest satisfaction for [area], Y% efficiency gain in [process], new revenue from [opportunity]).

- ______________________________
- ______________________________
- ______________________________

AFTERWORD

FROM PLAYBOOK TO PRACTICE: THE AI LITERACY MINDSET PROGRAM

The journey we have taken through this playbook, from understanding the fundamentals of AI to envisioning an AI-native hospitality future, is more than a theoretical exploration. The frameworks, strategies, and practical steps detailed within these pages are born from direct experience and the tangible results of our AI Literacy Mindset Program at Vivander Advisors. This playbook itself reflects that program's core teachings and successes.

Our AI Literacy Mindset Program is designed to cultivate lasting change in how hospitality professionals approach and utilize artificial intelligence. A key component of our methodology involves measurement and reinforcement. We begin with pre-program surveys to establish a baseline understanding of current

AI adoption levels, perceived skills, and attitudes toward AI within an organization. This data provides a crucial starting point.

Throughout the six-week intensive program, participants are immersed in building their AI literacy and actively guided in developing an AI-first mindset. To foster this, we implement daily AI tracker surveys. These serve as a consistent reminder for individuals to consciously consider AI in their daily tasks and to document at least one instance where they attempted to use AI, regardless of the outcome. This daily practice is fundamental to shifting from passive awareness to active engagement. Upon completion, post-program surveys allow us to measure the progress made. Consistently, these pre- and post-survey scores have demonstrated a significant increase in the adoption of advanced AI practices across the enterprise. Team members develop the beginnings of a true AI mindset, understanding how to leverage AI to solve problems and enhance their work strategically.

The results speak for themselves. Participants emerge from the program with a newfound confidence and a proactive approach to integrating AI. A common and exciting outcome we observe is that team members, having experienced the benefits of AI firsthand, quickly begin to identify further opportunities and are often the ones asking for more advanced solutions, including intelligent workflow automation.

This experience is something we at Vivander Advisors are passionate about replicating for others in the hospitality industry. If the principles and pathways outlined in this playbook resonate with you, and you are looking to cultivate a similar AI-first culture within your hotel, brand, or hotel management company, we can help. We are dedicated to guiding organizations like yours through their unique AI ascension. To take the next step and explore how we can help you scale AI effectively and responsibly in your organization, please contact us at vivanderadvisors.com. Let's build your AI-augmented future together.

www.ingramcontent.com/pod-product-compliance
Lightning Source LLC
Chambersburg PA
CBHW021918190326
41519CB00009B/827

* 9 7 9 8 8 9 0 7 9 3 4 2 3 *